Population Genetics

Population Genetics

Freddy B. Christiansen

Aarhus University

Marcus W. Feldman

Stanford University

Blackwell Scientific Publications, Inc.

Palo Alto, Oxford, London,
Edinburgh, Boston, Victoria

Editorial Offices
667 Lytton Avenue, Palo Alto, California 94301
Osney Mead, Oxford, OX2 0EL, UK
8 John Street, London WC1N 2ES, UK
23 Ainslie Place, Edinburgh, EH3 6AJ, UK
52 Beacon Street, Boston, Massachusetts 02108
107 Barry Street, Carlton, Victoria 3053, Australia

Distributors

USA and Canada
Blackwell Scientific Publications
P.O. Box 50009
Palo Alto, California 94303

UK
Blackwell Scientific Publications
Osney Mead
Oxford OX2 0EL

Australia
Blackwell Scientific Publications (Australia) Pty Ltd
107 Barry Street, Carlton
Victoria 3053

Library of Congress Cataloging-in-Publication Data
Christiansen, Freddy B., 1946–
Population genetics.

Bibliography; p.
Includes index.
1. Population genetics. I. Feldman, Marcus W.
II. Title. [DNLM: 1. Genetics, Population. QH 455 C555p]
QH455.C49 1986 575.1′5 85-15811
ISBN 0-86542-307-5

Manuscript editor: Victoria Nelson
Sponsoring editor: John Staples
Production coordination: Bookman Productions
Illustrator: Ginny Mickelson
Interior and cover design: Gary Head
Composition: Graphic Typesetting Service
Printing and binding: R. R. Donnelley & Sons

Preface

The essence of genetics is the study of variation among individuals that is transmitted from parent to offspring in some more or less reliable way. The causes, effects, mode of transmission, and change over time of this variation are all part of the science of genetics. Population genetics is the science of quantitative description of variation and rules of transmission in populations.

This book intends to provide a basic introduction to population genetics and to convey the spirit of interplay between observation and theory-aided interpretation that has characterized its history. It can be used either as a text for a short course in population genetics or as a supplementary text for a course in general genetics or population biology.

The variation that population genetics addresses can be due to the segregation of known Mendelian genes, it can be reliably transmitted and concentrated in families but without a known Mendelian origin, or it can be the result of a complex interaction between one or more genes and elements in the environment. For any particular trait, the population geneticist aims to describe variation and transmission in terms of some basic rules of genetic segregation and phenotypic determination, and to assess the importance of such evolutionary determinants as (1) differential mortality or fertility among the variants, (2) preferences among variants in mate choice, (3) admixture among subpopulations having different trait distributions, and (4) the effect of small population size. Each of these can be shown mathematically to influence the expectation among offspring of a given population of parents.

In population genetics, the use of models formulated in mathematical terms has played a significant and productive role. Such models help us to describe in detail the variation in natural populations and to make rigorous arguments concerning the origin and future of this variation. Above all, such models allow the variation to be described in terms of general genetic phenomena before it is possible directly to observe the operation of these phenomena in population studies.

The material we present here should be within reach of the student with some basic knowledge of classical genetics and high school mathematics, including very elementary probability. To describe variation in natural populations and to connect it with theory requires some statistical concepts and methods that we have attempted to keep simple; no formal training in statistics is needed to make use of them. The exercises have been chosen to augment the development in the text. A number of these problems introduce useful parts of the theory not covered in the text. Answers to a representative sample of these exercises are provided to help readers in gauging their progress.

Our presentation is oriented towards human population genetics and it is from humans that most of our examples are drawn. This emphasis reflects not only the general interest that most of us share in our own biology but also the central role played by population genetics in the development of human genetics. The other examples have been chosen both because of their relationship to general theory and because most of the organisms referred to are reasonably accessible in nature. Experimental animals such as fruitflies and mice have been central to the development of genetics and especially of population genetics, and they belong in any discussion of these subjects.

Many colleagues and students have knowingly or unwittingly contributed to this volume, and we thank them for their encouragement and inspiration. Drs. Ron Burton, Luca Cavalli-Sforza, Andrew Clarke, Rob Dorit, Kent Holsinger, Vibeke Simonsen, Volker Loeschke, and Richard Lewontin made useful criticisms of early drafts. We are indebted to Drs. Vivi Nielsen and Søren Andersen for permission to use Figures 4-2 and 4-3. The brilliant and cheerful word processing by Cheryl Nakashima simplified our task immensely.

FBC and MWF

Stanford, November 1984

Contents

1

Describing Variation in Populations

In every plant, animal, and human population, individuals are observed to differ from one another. This is *phenotypic* variation, and careful description allows us to address its causes in a scientific manner. Our interest here is in causes of variation that are peculiar to families.

The basic principle of genetics is that biological heredity is indirect. Genes, not phenotypes, are transmitted from parent to child, although the phenotype of an individual is influenced by its genotype. In describing phenotypic variation, therefore, we attempt to use methods that will reveal as much as possible about the genotype. It then becomes natural to classify phenotypic variation according to the degree of certainty with which the nature of its genetic causes, if any, may be inferred.

It is convenient to classify observed phenotypes into two groups. The first group consists of those characters that vary continuously along a scale of measurement. The individual is characterized by a *quantitative* trait value. This continuous variation describes a great many anthropometric variables, such as height, weight, IQ, plasma glucose fraction, and systolic blood pressure. In addition, such biochemical properties as the activity of a given enzyme vary continuously among different individuals of the same genotype at that enzyme locus. The trait values for each of these characters could in principle form a continuum, but such characters as the number of facets in the eye of an individual *Drosophila melanogaster,* or litter size in the laboratory mouse, which have a potentially large number of separate integral outcomes, are usually also included in this first class. The continuously varying characters and this class of discrete characters are combined into the class of *quantitative characters.*

The second class of variation includes those characters in which differences among individuals are *qualitative* rather than quantitative as in the first class. Included here would be the wrinkled versus smooth appearance of peas exploited by Mendel, and albino versus pigmented body of animals. Such dichotomies as normal versus harelip, clubfoot, cleft palate, diabetic, or schizophrenic conditions should also be included in this group even though the latter two, for example, may be difficult to define unambiguously. These examples are all dichotomous, but qualitative variation that allows complete classification into a few easily distinguished types would also be placed in the second group. This second class of varying characters exhibit *qualitative variation.*

It is important to differentiate the first two examples—wrinkled versus smooth peas, albino versus pigmented body—from the other qualitative traits. Each of the first group of characters satisfies Mendel's rules of genetic transmission; each is due to the operation of a single gene present in two or more allelic forms. Thus, each discrete phenotype is the expression of a simple genotype. This is therefore referred to as qualitative genetic variation. Although harelip, clubfoot, and the other diseases mentioned in the second group are to varying degrees concentrated in families, they do not segregate according to Mendelian rules. Indeed, their relationship to specific genes is unknown at this time. They are usually grouped with the quantitatively varying traits, primarily for theoretical reasons. We shall follow this custom and address the population properties of qualitative genetic variation separately.

1.1 Phenotypic Variation in Populations

Quantitative variation and qualitative variation without a clear Mendelian basis share the property that their transmission can only be revealed by a detailed study of phenotypic variation in the population. Such a study always involves a description of phenotypes, their distribution in populations, and the manner in which phenotypes aggregate within families. We will now provide a broad classification of phenotypic variation and examine some of the problems that arise in its descriptive analysis.

1.1.1 Quantitative Variation

Character variation in a population is described by means of a frequency distribution of the traits of individuals, that is, a graph of frequency in the population as a function of trait value. A *mode* of variation in the character is a local maximum or any peak in the trait frequency distribution in the population; if the frequency distribution exhibits a single mode, it is called *unimodal*. Many well-studied characters, including several of great interest to human genetics, are of this kind. Figures 1-1 through 1-4 illustrate a selection of unimodal quantitatively varying human traits. Figures 1-1 and 1-2 show the

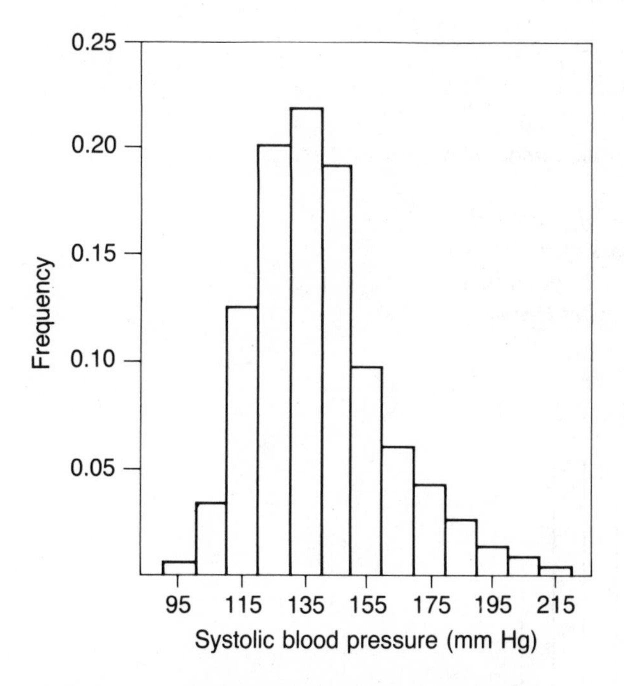

Figure 1-1. Distribution of systolic blood pressure (mm Hg) from the Framingham heart study. The observed distribution is given as a histogram showing the frequency of individuals with blood pressure in a given interval. For example, about 22 percent of the individuals had blood pressure between 130 and 139 mm Hg, and this is indicated as a column marked 135 mm Hg and height 0.22. The mean blood pressure is 137.7 mm Hg. (Courtesy Dr. Roger Williams.)

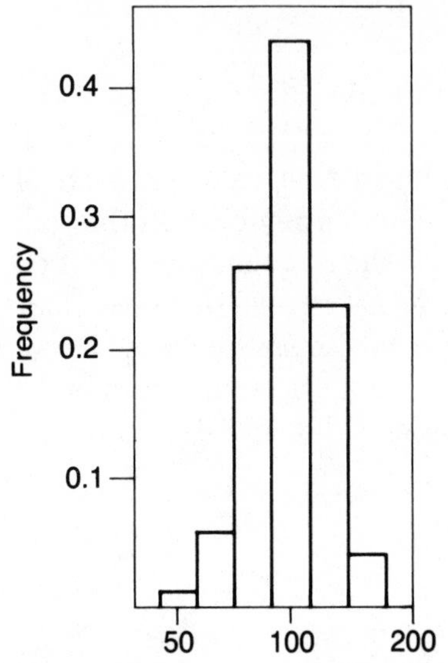

Figure 1-2. Distribution of plasma glucose concentrations, 2 hours after carbohydrate load, in boys ages 5–14 years in a study of Pima Indians. (After Bennett et al., 1976.)

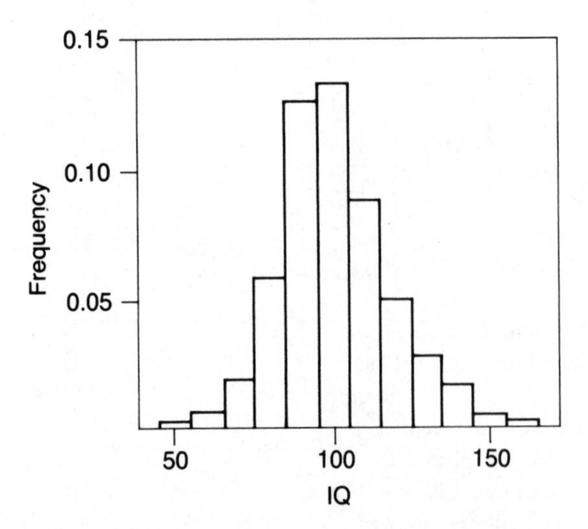

Figure 1-3. Distribution of a measure of IQ in Scottish children born in 1926. (After Mather, 1964.)

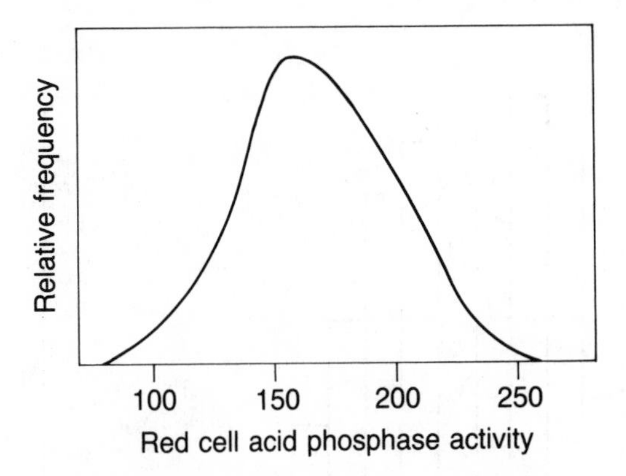

Figure 1-4. Distribution of red cell acid phosphatase activities among English adults. The distribution of data is given as a probability density function, where the frequency of individuals whose measurements fall within a given interval is obtained as the area under the curve above this interval. Activity is expressed as micromoles of p-nitrophenol liberated from p-nitrophenyl phosphate in 30 minutes at 37°C, per gram of hemoglobin present in hemolysate. (After Harris, 1966.)

variation among individuals in two routinely measured clinical characters: systolic blood pressure and plasma glucose concentration. The variation in physiological characters like these is due to three factors, namely, random errors in the clinical measurement of the trait, day-to-day variation in the value assigned to the individual, and finally variation among individuals in their characteristic blood pressure level or glucose concentration, respectively. Our interest here, of course, is the last source of trait variation in the population, namely, the true variation among individuals. The first two sources of variation may be viewed as merely statistical errors. In the investigation of continuously varying traits, however, individual measurements are always prone to statistical errors. Although this error may be decreased by repeated measurements of each individual, we must accept this inherent uncertainty in the determination of the trait of the individual.

Figures 1-5 through 1-8 are taken from experimental studies of mouse and *Drosophila* populations. Figure 1-5 shows the variation in a morphometric character of a type frequently used for experimental investigations of quantitative genetics. The trait (number of bristles)

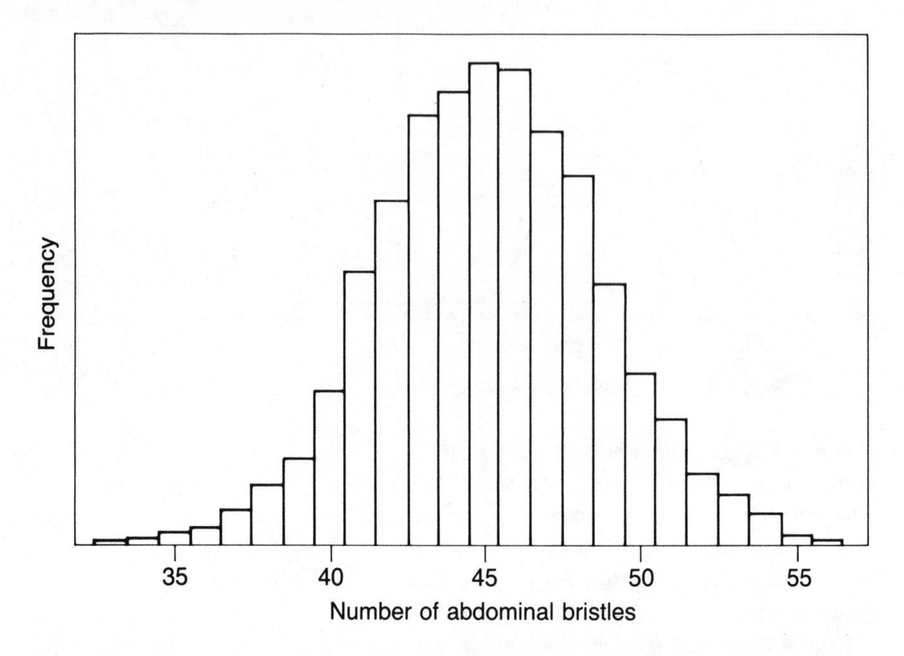

Figure 1-5. Distribution of abdominal bristle number in a population of fruit-flies, Drosophila melanogaster. *The histogram shows the frequency among 3264 flies with the given number of bristles. The mean bristle number is 45.15 and the standard deviation of bristle numbers in the population is 3.51. (After data of Sheridan et al., 1968.)*

is clearly integer valued, but a comparison of the variation in this and the characters discussed earlier shows the superficial similarity in the population distribution that justifies the inclusion of this kind of discrete character in the same class as the continuously varying characters.

The two characters of mice described in Figures 1-7 and 1-8 are related to yield characters of livestock, and they have been widely used for model experiments in animal breeding. The growth rate of mice in Figure 1-8 is an example of a character that is heavily dependent on the environment in which the mice are kept. Mice that are fed reduced quantities or receive feed of low nutritional value will obviously grow more slowly than those that receive ample feed of high nutritional value. Thus characters like this may be used to investigate the influence of the environment on the phenotype and on the relation between the phenotype and the genotype of the individual.

In the further study of unimodal continuous traits it is often assumed that the frequency distribution is normal (Figure 1-9); the mean m and variance V are sufficient to describe it statistically. This distri-

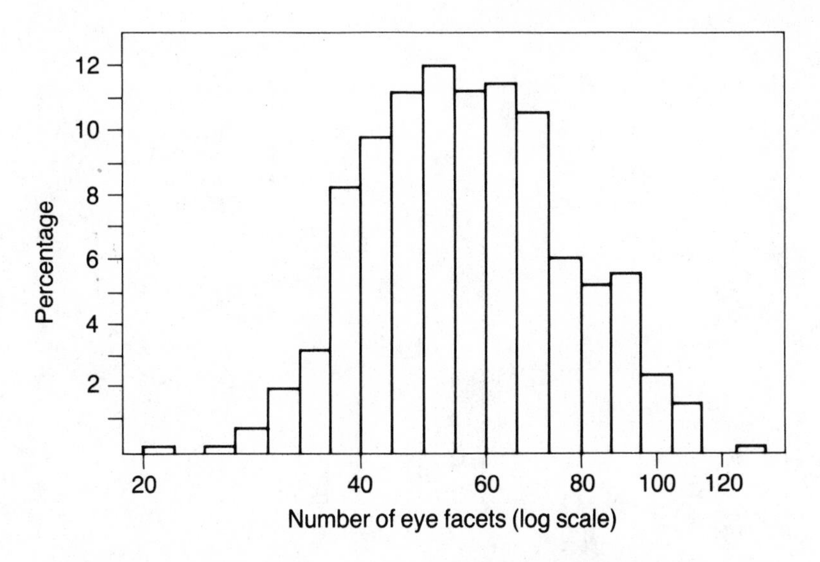

Figure 1-6. Distribution of eye facet number in females of a Drosophila melanogaster *population homozygous for the eye mutant* Bar. *(After Falconer, 1981; data from Zeleny, 1922.)*

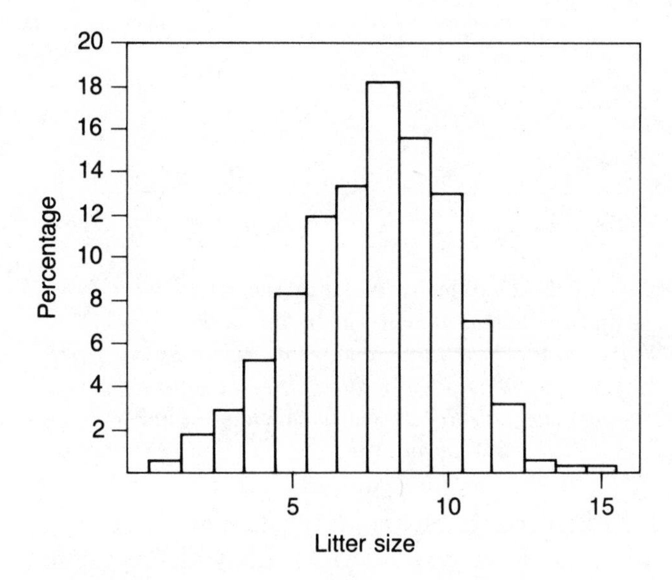

Figure 1-7. Distribution of litter sizes in mice. (After Falconer, 1981.)

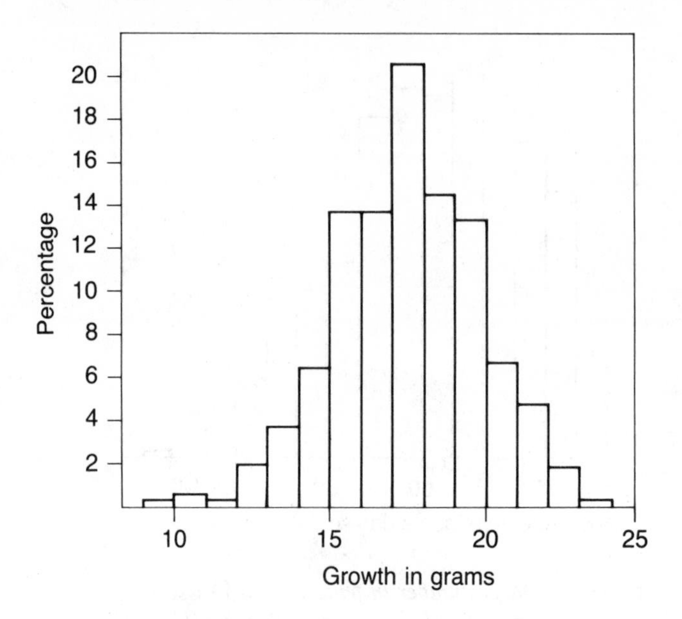

Figure 1-8. Distribution of growth in male mice ages 3 to 6 weeks. (After Falconer, 1981.)

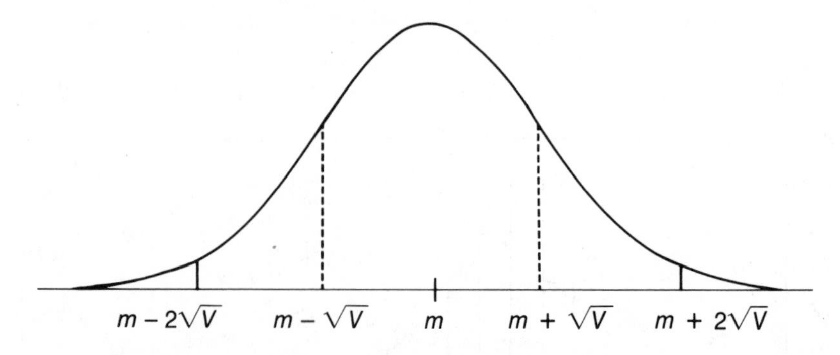

Figure 1-9. The probability density of the normal distribution with mean m *and variance* V. *The probability that an observation falls in the interval between* $m - \sqrt{V}$ *and* $m + \sqrt{V}$ *is 0.683, and the interval from* $m - 2\sqrt{V}$ *to* $m + 2\sqrt{V}$ *has a probability of 0.955. Thus, the standard deviation,* $\sqrt{V}$, *provides a handy guide to expected deviation of the individual measurement from the theoretical mean* m.

bution assumption is often reasonable, as in the case of Figure 1-3, but in Figure 1-1, for example, there is a lack of the symmetry about the mean that is characteristic of the normal distribution. Often the use of logarithmic scale decreases or removes this asymmetry.

Figure 1-10 is an example of variation in a discrete quantitative character in natural populations. The form of the trait distribution (as described by the modality and the variance) is rather similar in the three populations, but the mean trait value varies geographically among populations.

Exercise 1.1.A

The population sample from location 2 in Figure 1-10 follows.

Vertebrae number	Number of fish observed
104	1
105	2
106	10
107	12
108	25
109	40
110	35
111	32
112	11
113	5
114	2
115	2
Total	177

From this sample the population mean can be estimated by

$$\hat{m} = \frac{\sum_{j=104}^{115} jN_j}{177},$$

where N_j is the number of fish observed to have j vertebrae.

a. Check that $\hat{m} = 109.46$.

The phenotypic variance in the population can be estimated by

$$\hat{V} = \sum_{j=104}^{115} \frac{(j - \hat{m})^2 N_j}{176}$$

b. Check that $\hat{V} = 3.67$. ($\sqrt{\hat{V}} = 1.92$)

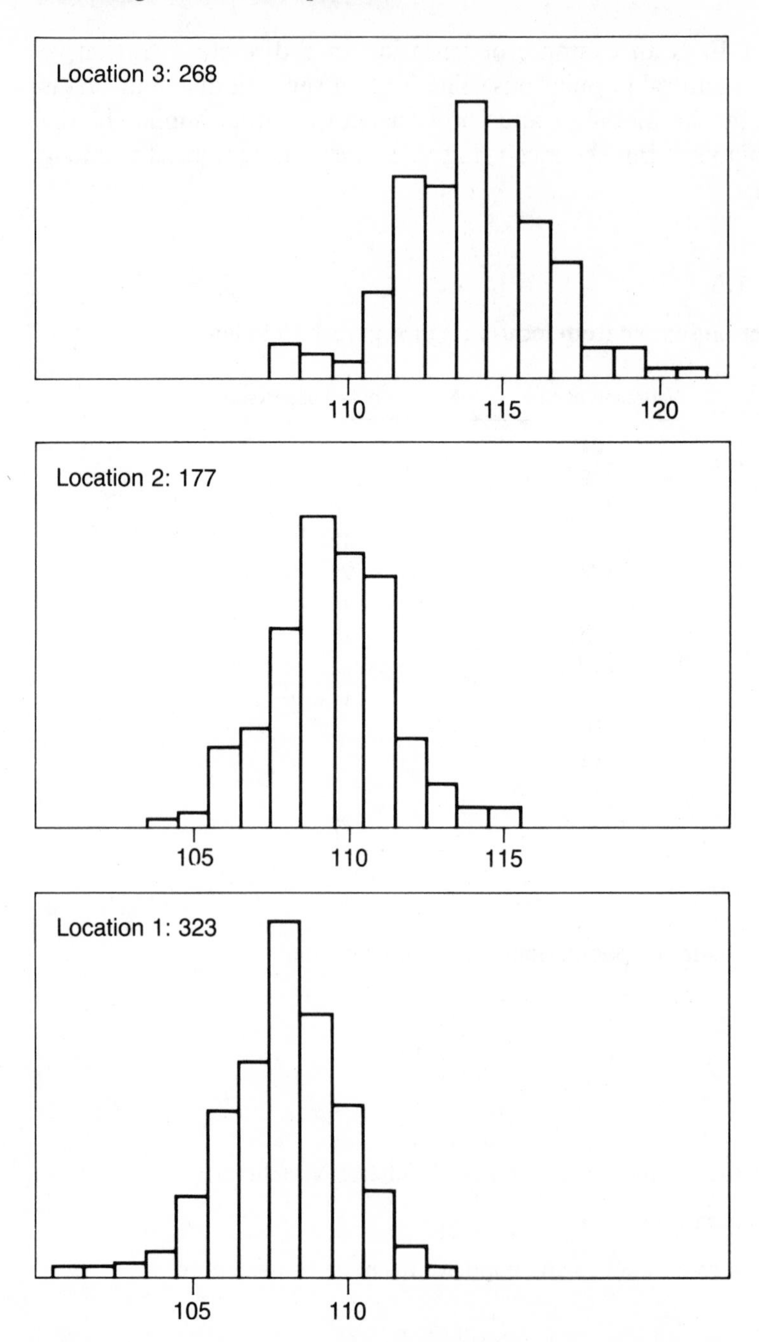

Figure 1-10A. Geographical variation within a Danish fjord in the population distribution of the number of vertebrae in the marine fish Zoarces viviparus. *(After Schmidt, 1917.)*

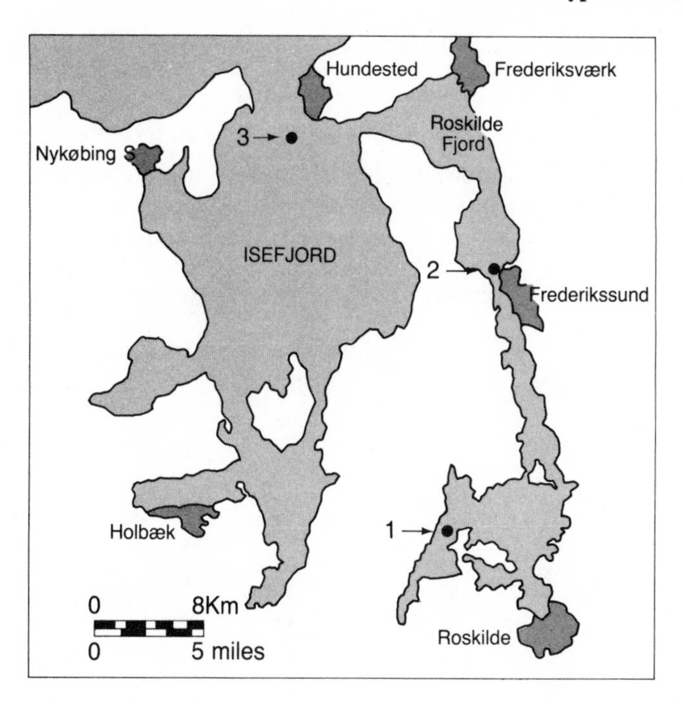

Figure 1-10B. Locations 1, 2, 3 in Figure 1-10A.

If the variation in vertebrae number can be described by a normal distribution, about 70 percent of the observations are expected to lie between $m - \sqrt{V}$ and $m + \sqrt{V}$, and about 95 percent between $m - 2\sqrt{V}$ and $m + 2\sqrt{V}$ (Figure 1-9).

c. What proportion of the fish are *observed* to lie in the interval between $\hat{m} - \sqrt{\hat{V}}$ and $\hat{m} + \sqrt{\hat{V}}$, and between $\hat{m} - 2\sqrt{\hat{V}}$ and $\hat{m} + 2\sqrt{\hat{V}}$?

For a continuously varying character, similar expressions can be derived. Just group the observations in a reasonable number of equal-sized intervals (usually about 20). If there are n such intervals numbered $j = 1,2, \ldots ,n$, let x_j be the *midpoint* of the jth interval in which N_j observations lie. Then

$$\hat{m} = \sum_{j=1}^{n} \frac{x_j N_j}{\sum_{j=1}^{n} N_j}$$

$$\hat{V} = \sum_{j=1}^{n} \frac{(x_j - \hat{m})^2 N_j}{\sum_{j=1}^{n} N_j - 1}$$

d. Convince yourself that these are the same expressions as would be used for a discrete character.

Exercise 1.1.B

The population samples from locations 1 and 3 in Figure 1-10 are shown in the following table. Calculate $\hat{m}$ and $\hat{V}$ for each of the two populations.

Vertebrae number	Location 1	Location 3
101	2	
102	2	
103	3	
104	5	
105	18	
106	35	
107	48	
108	83	5
109	61	4
110	39	3
111	20	16
112	6	39
113	1	37
114		53
115		49
116		29
117		21
118		5
119		5
120		1
121		1
Total	323	268

A frequency distribution of a phenotype in a population is called *multimodal* if it has more than one peak. This is not a very common observation for quantitative characters. Moreover, there is considerable discussion in the statistical literature about when an observed frequency distribution may reasonably be inferred to represent a simple form, for example, a truly bimodal distribution. A multimodal character may be viewed as an intermediate between the common appearance of a unimodal quantitative character and a qualitative character. Indeed, minor variation among individuals exhibiting a given qualitative character is often well described as a *unimodal* quantitative trait distribution (Figures 1-6 and 1-11). If we imagine the variation

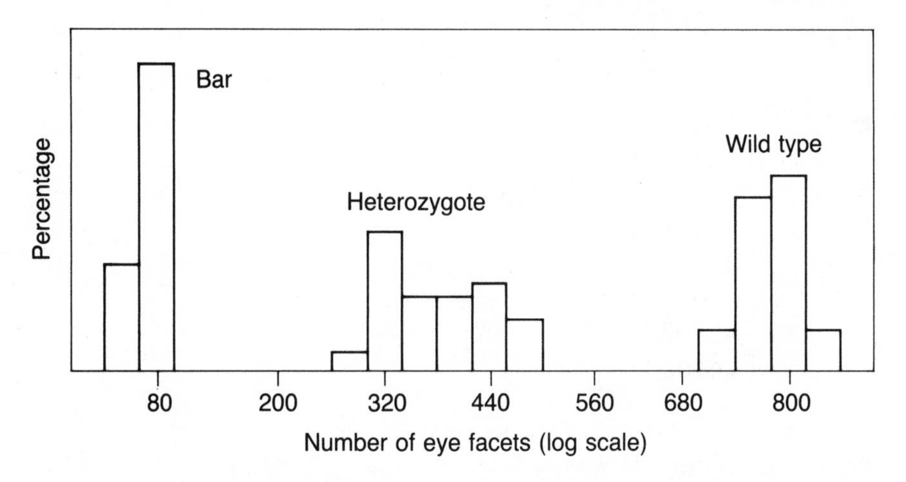

Figure 1-11. The difference in number of eye facets among the three genotypes at the Bar *locus in* Drosophila melanogaster *females. The three genotypes have qualitatively distinct and easily recognizable phenotypes, but the number of eye facets varies within each phenotypic class (see also Figure 1-6). (After Sturtevant and Beadle, 1962.)*

among individuals within qualitative trait classes to increase, then we reach a point where unambiguous qualitative differences among individuals no longer exist, and the distribution of trait values is best described as a multimodal frequency distribution.

In genetics, bimodality of the phenotype distribution is often taken as suggestive of two genotypic classes, one for each mode. But most cases are not as clear cut as Figure 1-11. Overlap can be seen in the plasma glucose levels of Pima Indians aged 45–54 years illustrated in Figure 1-12. Compare this with the 5–14 year olds in the same population, who were truly unimodal on the same scale (Figure 1-2). This age dependence is an important factor for many diseases and must be considered in comparing phenotype distributions among populations.

1.1.2 Discrete Non-Mendelian Traits

Most of the common congenital malformations such as clubfoot, anencephaly, and pyloric stenosis are "all-or-none" traits. They are not, however, transmitted according to Mendel's rules. The data of Carter (1965) suggest that the population incidence of such diseases (excluding known chromosomal abnormalities) reaches 1–2 percent. Excluded from this figure are two other important diseases whose diagnoses are much less clear cut, namely, diabetes mellitus and schizophrenia. The incidence of diabetes among people less than 25 years old is about 0.1

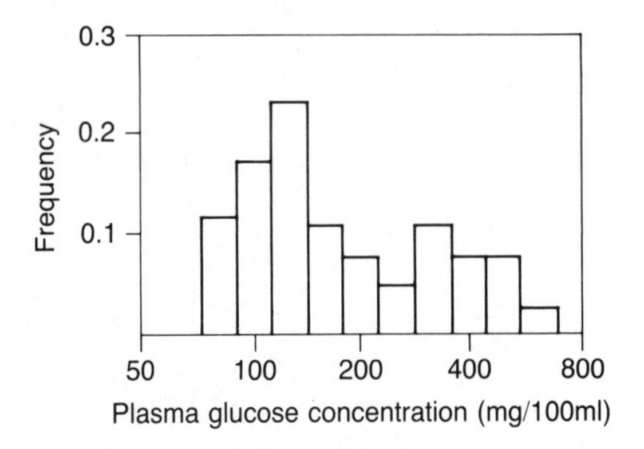

Figure 1-12. Distribution of glucose concentration 2 hours after carbohydrate load in adult males ages 45–54 in the study of Pima Indians cited in Figure 1-2. (After Bennett et al., 1976.)

percent, but in the 60–70-year age group the frequency may reach 3–6 percent. The age of onset varies widely, with the more severe cases, so-called juvenile onset diabetes, occurring before age 20. Other criteria, to be discussed later, support the idea that the early onset form is a separate entity. Although schizophrenia is a difficult disease to diagnose, the main symptoms are generally accepted to be anomalous thought patterns, loss of contact with reality, or a generally inappropriate response to external stimuli. The population incidence is about 1 percent in the United States.

Although the total incidence of these discrete diseases reaches over 8 percent in most of the United States, the frequency in some subpopulations is much higher. In the Pima Indians, for example, 47 percent of males aged 65–74, and 69 percent of females aged 55–64 suffer from diabetes (Bennett et al., 1976).

A framework often used in the discussion of non-Mendelian traits, whose definition involves a sharp qualitative difference, involves the concept of liability. The assumption is made that underlying the observed trait difference is continuous, unimodal (usually normally distributed) variation in a trait, called *liability* (Figure 1-13; Falconer, 1965). If the value of this liability is larger than some value, called the *threshold*, then the first trait—namely, the disease—is manifested. Otherwise the individual is normal. The formal demonstration of the existence of such a liability function has yet to be accomplished. In diabetes mellitus, however, tolerance to glucose loading, which has the prop-

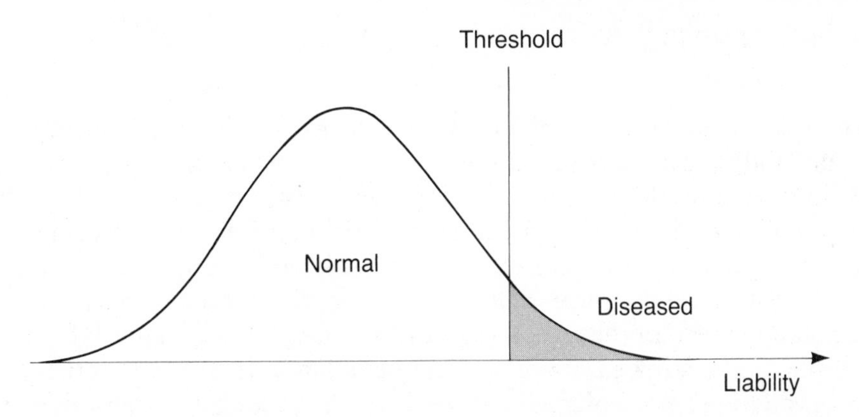

Figure 1-13. Theoretical model of a disease. Individuals in the population are assumed to vary with respect to one character: Liability to the disease. Individuals with a liability above a threshold value express the disease, whereas individuals with a liability below the threshold are normal.

erties of a liability function described earlier, has been used as a liability trait. Other criteria such as blood sugar level or insulin activity, while analogues of liability variables, are usually the disease-defining criteria. In combination with other assumptions about its genetic basis, to be discussed later, liability has been used as a basis for theoretical predictions about the risk of disease in families.

1.1.3 Familial Aggregation

Most of the traits described so far show *familial aggregation,* that is, related individuals will show on the average greater resemblance for the trait than unrelated individuals. For the diseases and malformations referred to in section 1.1.2, familial aggregation entails that individuals related to an affected individual, called the *proband,* have a higher risk of acquiring the disease or have a trait value closer to the proband than to a randomly chosen member of the population. In general, familial aggregation for a trait is present if prior knowledge of the phenotypic value of a family member alters the probability distribution of phenotypes among the other members of the family. Phenotypes that are completely determined by a single Mendelian gene will surely show familial aggregation, but for most of the traits described so far no simple genetic etiology has been discovered despite their familial aggregation.

1.1.3.1 Quantitative Variation

The most obvious example of familial aggregation, or correlation among related individuals, is the resemblance between parents and offspring, as illustrated by the mother-offspring resemblance described in Figures 1-14 and 1-15. The phenotypic distribution among offspring resembles that among mothers (Figure 1-14), and for mothers of a given trait value the mean value of the offspring deviates from the population mean in the direction of the mother's value (Figure 1-15). Despite the striking parent-offspring resemblance, there is still rather large variation among offspring of mothers who have a given trait value (Figure 1-16). This residual variation among offspring is due partly to their resemblance to the unknown father, partly to genetic segregation, and partly to the influence of environmental variation.

A variety of tools has been used to compare levels of familial aggregation. To some extent the utility of these statistics depends on the type of trait. For variation that is unimodal and approximately normal, such as systolic and diastolic blood pressure, height, and weight, the *product-moment correlation* is often used. For n pairs of related individuals with phenotypic values X_{1i} and X_{2i}, $i = 1,2, \ldots, n$, this quantity is given by

$$r = \sum_{i=1}^{n} \frac{(X_{1i} - m)(X_{2i} - m)}{nV}$$

where m is the mean and V is the variance of the phenotypic distribution in the population. The correlation coefficient is restricted to the interval $-1 \leqslant r \leqslant 1$. The value $r = 0$ corresponds to no relationship, and $r = 1$ corresponds to $X_{1i} = X_{2i}$, that is, perfect agreement. An example of such a comparison among biologically related individuals with respect to the four traits just mentioned, using data from Muscatine, Iowa, is presented in Table 1-1. As we shall see later, compilations of correlations among individuals of various degrees of relationship (such as those in Table 1-1) are used in the construction of models for the etiology of the trait.

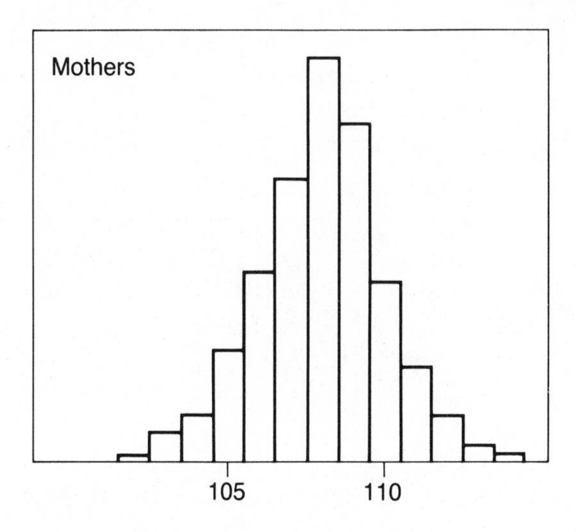

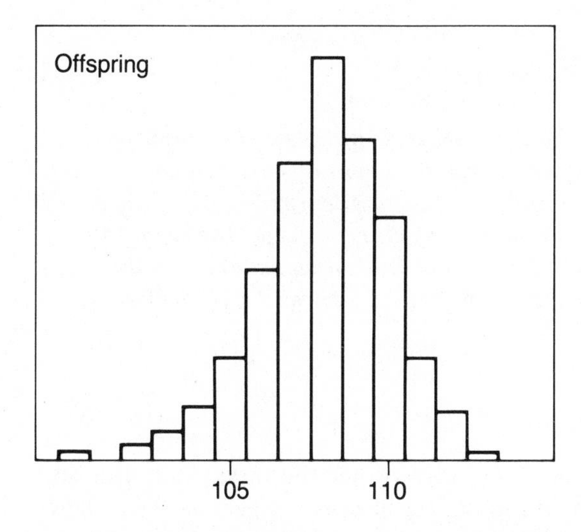

Figure 1-14. Comparison between generations of the population distribution of the number of vertebrae in Zoarces viviparus, *a live-bearing fish. Top histogram: vertebrae counts in 631 pregnant mothers. Bottom histogram: vertebrae counts in 3155 of their foeti (five from each mother). (Data from Smith, 1922.)*

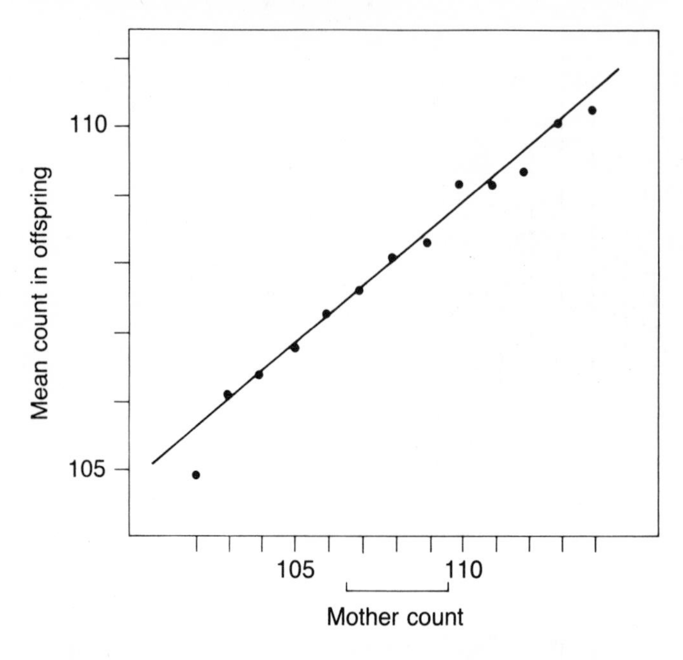

Figure 1-15. Comparison of vertebrae counts in mothers and offspring of Zoarces viviparus. *For mothers with a given number of vertebrae, the mean vertebrae count of their foeti (five from each mother) is determined, and the figure shows the offspring mean as a function of the mother's value. The mother-offspring similarity may be described as a linear relationship indicated by the line of slope 0.4 (the unit on the ordinate axis is twice as large as the unit on the abscissa). (After Smith, 1922.)*

Exercise 1.1.C

Schmidt (1917) and Smith (1922) also investigated the variation in the number of rays in the pectoral fin in populations of *Zoarces viviparus*. The following table shows the ray counts in 1760 mothers and in one randomly chosen offspring for each mother.

	Mother count				
Offspring count	18	19	20	21	Total
17	0	1	0	0	1
18	34	147	22	2	205
19	87	768	310	12	1177
20	9	173	172	13	367
21	0	1	6	3	10
Total	130	1090	510	30	1760

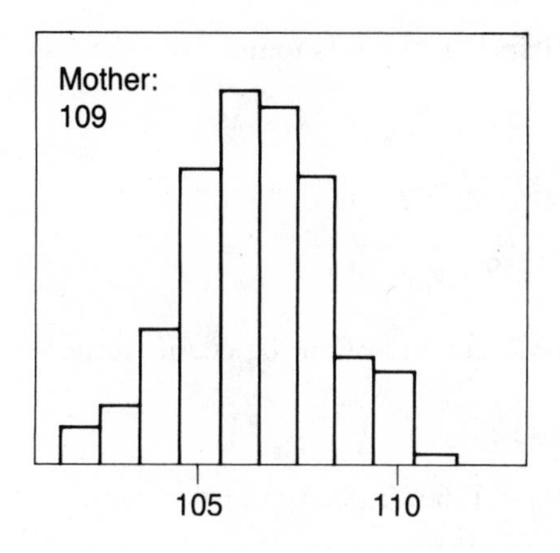

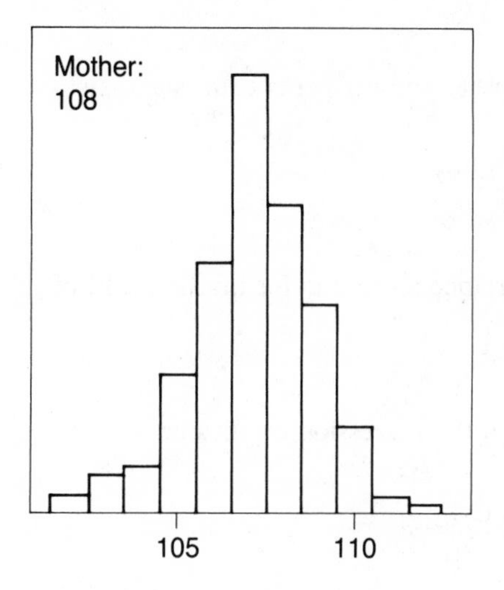

Figure 1-16. The variation among offspring from a mother with a given vertebrae count in Zoarces viviparus. *Variation among offspring is shown for the three most common values among the mothers (Figure 1-14 and bracketed in Figure 1-15). (Data from Smith, 1922.)*

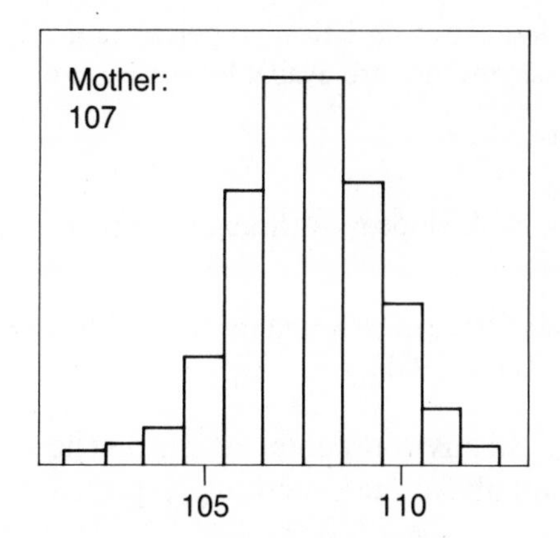

a. Draw a figure like Figure 1-15 using these observations.

The correlation between mother and offspring values is found from the covariance estimate

$$\hat{C}_{MO} = \frac{\sum_{i=18}^{21} \sum_{j=17}^{21} N_{ij}(i - \hat{m}_M)(j - \hat{m}_O)}{1759},$$

where $\hat{m}_M$ and $\hat{m}_O$ are the means of the mother and offspring counts respectively. Alternatively,

$$\hat{C}_{MO} = \frac{\sum_{i=18}^{21} \sum_{j=17}^{21} ijN_{ij} - 1760\,(\hat{m}_M\hat{m}_O)}{1759}.$$

The correlation coefficient between mother and offspring counts is then

$$\hat{r}_{MO} = \frac{\hat{C}_{MO}}{(\hat{V}_M\hat{V}_O)^{1/2}}$$

where $\hat{V}_M$ and $\hat{V}_O$ are the estimated variance in counts for mothers and offspring, respectively (see Exercise 1.1.A).

b. Calculate $\hat{r}_{MO}$.

To evaluate mother-offspring similarity, the regression coefficient

$$\hat{b}_{MO} = \frac{\hat{C}_{MO}}{\hat{V}_M}$$

is often used (we expect $V_O = V_M$), as it has better statistical properties (e.g., it can be easily used if more offspring per mother are available, see Figure 1-16).

c. Calculate $\hat{b}_{MO}$.

d. In the drawing for (a), place the line with slope $\hat{b}_{MO}$ through the point $(\hat{m}_M, \hat{m}_O)$.

Further discussion of these statistical quantities can be found in any textbook on biostatistics, for example, Sokal and Rohlf (1981).

Relatives often share environments, so environmental effects on the trait may inflate the familial aggregation above the genetically expected

Table 1-1. *Correlation Coefficients of Phenotypic Values Between Related Individuals.* One pair of individuals is selected at random from each investigated family. The number of pairs studied is shown in parentheses after each value of the correlation coefficient. If there is no familial aggregation, the correlation coefficients are expected to be 0, and all the correlation coefficients were compared to this value by a statistical test. The notes indicate the level of significance of this test.

Relationship	Systolic blood pressure	Diastolic blood pressure	Height	Weight
Sib-sib	0.12(119)	0.30(119)[c]	0.32(128)[c]	0.27(128)[b]
Parent-child	0.09(128)	0.21(128)[a]	0.27(131)[b]	0.29(131)[b]
Grandparent-grandchild	0.08 (69)	0.16 (69)	−0.06(110)	0.03(110)
Aunt/uncle-niece/nephew	0.19 (68)	−0.09 (68)	0.25(120)[b]	0.13(115)
First cousin	−0.17 (60)	0.01 (60)	0.01(114)	0.12(113)
Mother-father	0.01(119)	0.00(119)	−0.05(119)	0.04(119)

After Schrott (1979). a. $p < 0.05$, b. $p < 0.01$, c. $p < 0.001$.

level. For humans, as well as other mammals, siblings share an important environment in early life: their mother. This possible *maternal effect* on a trait may be revealed by, for instance, comparing the resemblance of halfsibs sharing the mother to the resemblance of halfsibs sharing the father. Alternative ways to uncover maternal effects or other effects of common environments are discussed later, together with the theory of inheritance of quantitative characters. Most of these methods, however, require observations in manipulated experimental populations, so they are not applicable to the study of human populations.

The closest degree of genetic relationship occurs between monozygous (MZ) twins. Because they are genetically identical, these special individuals seem ideal for the evaluation of the range of phenotypes that a given genotype can express. Monozygous twins, however, also experience a unique environment that cautions against an uncritical interpretation of phenotypic comparisons of twin pairs. Nevertheless, monozygous twins have received attention from two points of view. First, if they are separated at or shortly after birth, and reared apart, then the two "familial" contributions to any trait from the biological and adoptive parents are potentially separable. Second, monozygous

Table 1-2. *Intraclass Correlation Coefficients of Phenotypic Values in Twins.*

Character	MZ twins	DZ twins
Fingerprint ridge count	.96	.47
Height	.94	.44
Weight	.92	.63
IQ	.76	.51

Data on fingerprint ridge count from Huntley (1966) and on height, weight, and IQ from Shields (1962).

twins are compared to same-sex dizygous (DZ) twins who are genetically the same as ordinary sibs, but who, because of their age similarity, might be expected to interact more like monozygous twins. Similarity between twins is often measured using an intraclass correlation coefficient, which is

$$\rho = 1 - \sum_{i=1}^{n} \frac{(X_{i1} - X_{i2})^2}{2nV'},$$

where X_{i1} and X_{i2} are the phenotypic values of a pair of twins and V' is the phenotypic variance of twins. The intraclass correlation coefficient is the same as r if $V' = V$, and like r it is restricted to the interval $-1 \leq \rho \leq 1$. Examples for IQ, height, weight, and mean ridge count in fingerprints are presented in Table 1-2.

In the assessment of levels of familial aggregation, as in Tables 1-1 and 1-2, there are numerous statistical subtleties that cannot be ignored. Perhaps the most important of these is related to subgroup differences in the trait within the overall population. For example, many of the traits in Tables 1-1 and 1-2 have different distributions in males and females, in different age classes, or among different socioeconomic classes. The techniques that correct for these differences are called *standardization*. A particularly comprehensive analysis of this process with respect to blood pressure is described by Schork et al. (1977). Systematic age differences in such traits as blood pressure, plasma cholesterol, and blood sugar levels have been found. Hypertension, chronic heart disease, and diabetes are among the most important common diseases (see, for example, Sing and Skolnick, 1979), and techniques of standardization play a central role in modern research into their familial aggregation.

The need for standardization is illustrated by Table 1-3, where we see that in the Detroit study males and blacks were more hypertensive than females and whites, respectively (see also Feinleib and Garrison, 1979). A major issue is what variables, apart from the one under study, should be subject to standardization. For example, besides age and sex, weight, overall diet, and sodium intake might also affect blood pressure, and each of these variables also shows familial aggregation.

An important element (not included in the family aggregation Tables 1-1 and 1-2) that is essential for a reasonable interpretation of such data is the correlation between spouses' phenotypic values. This correlation is called the degree of *assortative mating*. The husband-wife correlation for IQ has been observed to be about 0.5 in the United States (Rao et al., 1976), and the correlation between spouses for educational attainment is very strong (Cavalli-Sforza and Bodmer, 1971, p. 548). Undoubtedly these traits influence the child's educational attainment, so if our ultimate interest is in predicting the educational attainment of children, then the appropriate variable to report for the parent is open to question. Apart from the problem of variable definition, a fundamental limitation of measured assortative mating is that the measurement is usually made in conjunction with a similar measurement on a child. Thus the correlation is not measured at the time of marriage, and the traits are often likely to have changed in the direction of increased correlation from the time of marriage to the time of measurement. At present there is no acceptable way to cope with this process of *marital convergence*.

1.1.3.2 Genetics of Quantitative Traits

Parent-offspring resemblance (Figure 1-15) and other familial aggregations of quantitative traits appear to agree with an intuitive picture of the direct inheritance of such attributes as wealth. However, they seem to differ from Mendelian inheritance, which is indirect, in that an offspring inherits its genes, not its phenotype, from the parents. This apparent discrepancy is resolved by viewing a quantitative character as influenced by many individual genes, each giving a small contribution to the genotypic determination of the trait (Figure 1-17). As we shall see in Chapter 4, this model, originally formulated by R. A. Fisher in 1918, provides a good framework for the description of many characters, and experimental support of the model accumulated around the same time as it was proposed (for example, from the investigations of Nilsson-Ehle, 1909). From the studies of monozygous twins it is evident that two individuals of the same genotype do not have the same phenotype. Such variation between individuals is ascribed

Table 1-3. *Distribution of Systolic and Diastolic Blood Pressure Categories by Sex and Race.* Measurements on 1844 individuals of systolic and diastolic blood pressure grouped into three groups: normal blood pressure, borderline blood pressure, and hypertensive blood pressure. The 1844 individuals are separated into classes according to sex and race, providing the "race-sex" table on the

	Sex				Race			
	Male		Female		Black		White	
Blood pressure	%	(*N*)	%	(*N*)	%	(*N*)	%	(*N*)
Systolic								
Normal ⩽ 139 mm	78	(720)	79	(728)	77	(701)	81	(747)
Borderline 140–159 mm	17	(154)	14	(132)	16	(146)	15	(140)
Hypertensive ⩾ 160 mm	6	(53)	6	(57)	8	(69)	4	(41)
Total	100	(927)	100	(917)	100	(916)	100	(928)
	$\chi^2_{(2)} = 1.83$				$\chi^2_{(2)} = 8.71$[b]			
Diastolic								
Normal ⩽ 89 mm	73	(674)	78	(722)	71	(650)	80	(746)
Borderline 90–94 mm	11	(102)	11	(101)	13	(120)	9	(83)
Hypertensive ⩾ 95 mm	16	(151)	10	(94)	16	(146)	11	(99)
Total	100	(927)	100	(917)	100	(916)	100	(928)
	$\chi^2_{(2)} = 14.98$[d]				$\chi^2_{(2)} = 22.38$[d]			

Data from Harburg et al. (1977). After Vogel and Motulsky (1978).
a. The interaction $\chi^2_{(2)}$ tests the hypothesis that the information in the race-sex column is properly represented by the two simple race and sex columns to the left.
b. $P < 0.05$.
c. $P < 0.01$.
d. $P < 0.001$.

to variations in the environment of the individuals, and it is called the *environmental variation.* This distinction between the effect of the genotype and the effect of the environment on the phenotype was originally made by W. Johannsen (1903) in a study of beans, where, because of experimental inbreeding, individuals of the same genotype are more easily available than in humans.

Those quantitative traits which satisfy this description of their inheritance are said to show *polygenic* or *multifactorial* inheritance.

right. Summing over race and sex, respectively, provides the “sex” and “race” tables on the left. For each column the distribution of the individuals on the three blood pressure groups is given in percent (and as number of individuals in parenthesis). For each of the tables a χ^2-test is given for equal blood pressure distribution among individuals of different classes.[a]

Race-sex							
Black male		White male		Black female		White female	
%	(*N*)	%	(*N*)	%	(*N*)	%	(*N*)
77	(350)	79	(370)	77	(351)	82	(377)
16	(74)	17	(80)	16	(72)	13	(60)
7	(33)	4	(20)	8	(36)	5	(21)
100	(457)	100	(470)	100	(470)	100	(458)
$\chi^2_{(6)} = 11.67 = 1.83 + 8.71 + 1.13$ Interaction $\chi^2_{(2)} = 1.13$							
69	(317)	76	(357)	73	(333)	85	(389)
13	(59)	9	(43)	13	(61)	9	(40)
18	(81)	15	(70)	14	(65)	6	(29)
100	(457)	100	(470)	100	(459)	100	(458)
$\chi^2_{(6)} = 43.39^d = 14.98 + 22.38 + 5.68$ Interaction $\chi^2_{(2)} = 5.68$							

1.1.3.3 Qualitative Traits

For qualitative traits, correlation coefficients are not commonly used to describe familial aggregation. There are many other descriptive statistical methods in common use, only one of which will be mentioned here. If all other variables (that is, environmental factors) were controlled, then the proportion of relatives of an affected individual who are also affected, compared to the incidence of the trait in the population at large, should give some indication of the importance of genetics in determining the trait. In practice, environmental variables are not subject to control, and for rare traits difficulties in diagnosis may inflate familial aggregation. Familial incidence, however, is often

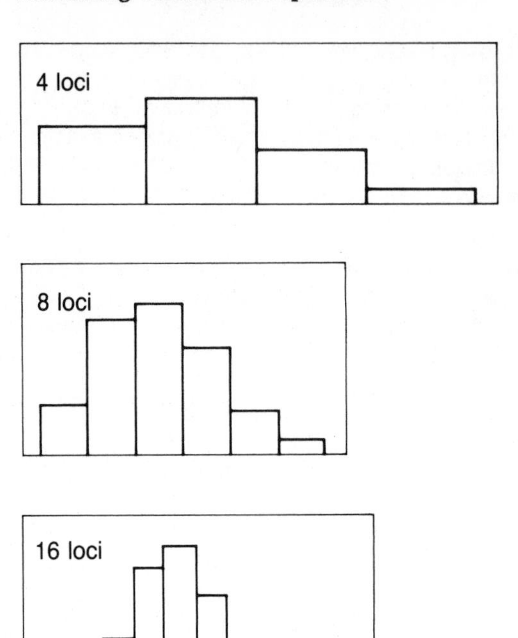

Figure 1-17. Theoretical phenotypic distributions in the model of polygenic inheritance. Top histogram: phenotypic distribution in the F_2 *generation after the cross AABBCCDD* × *aabbccdd, where big letters correspond to dominate alleles, small letters to recessive alleles. The phenotype of an individual is defined as the number of loci where it has the recessive genotype (e.g., Aa BB cc dd has a phenotypic value of 2, the* F_1 *individuals, AaBbCcDd, a value of 0, and the two individuals in the above cross have the phenotypes 0 and 4, respectively). Distribution among individuals in* F_2 *is obtained by assuming independent segregation at the four loci (Mendel's second law). Distributions for 8 and 16 loci can be obtained using a similar model.*

useful for empirical prediction and subclassification of diseases. A widely used incidence measure is the *concordance* among monozygous twins, namely, the fraction of pairs both of whom have the trait relative to all pairs of monozygous twins at least one of whom is affected. Obviously, if the trait were entirely under genetic control, the concordance would be 100 percent. In addition, the concordance among monozygous twins would be expected to be higher than among dizy-

Table 1-4. *Twin Concordance in Multifactorial Diseases.* The number *N* and percent of concordant pairs in the two types of twins is shown for a number of morphological qualitative traits and diseases (excluding mental diseases).

Condition	Twin type	Number	Concordant *N*	Concordant %	MZ more frequently concordant than DZ
Clubfoot	MZ	35	8	22.9	10.0 times
	DZ	135	3	2.3	
Congenital dislocation of hip	MZ	29	12	41.4	14.8 times
	DZ	109	3	2.8	
Cleft lip, palate	MZ	125	37	29.6	6.4 times
	DZ	236	11	4.7	
Cancer	MZ	196	34	17.4	1.6 times
	DZ	546	59	10.8	
Coronary heart disease	MZ	21	4	19.0	2.4 times
	DZ	47	4	8.5	
Diabetes mellitus	MZ	181	101	55.8	4.9 times
	DZ	394	45	11.4	
Atopic diseases	MZ	12	6	50.0	11.0 times
	DZ	23	1	4.4	
Hyper-thyroidism	MZ	49	23	47.0	15.1 times
	DZ	64	2	3.1	
Psoriasis	MZ	31	19	61.0	4.7 times
	DZ	46	6	13.0	
Cholelithiasis	MZ	49	13	26.6	4.1 times
	DZ	62	4	6.5	
Tuberculosis	MZ	381	202	51.6	2.3 times
	DZ	843	187	22.2	
Sarcoidosis	MZ	4	2	50.0	5.9 times
	DZ	11	1	8.5	

After Vogel and Motulsky (1979).

Table 1-5. *Familial Aggregation in Multifactorial Diseases.* Proportion of affected relatives of persons with some common malformations is given as a factor times the incidence in the general population (first row).

	Harelip (+ cleft palate)	Congenital dislocation of the hip		*Talipes equinovarus*	Pyloric Stenosis		Anencephalus and spina bifida
		All patients	Female relatives of female patients		All patients	Female relatives of female patients	
Population incidence (approximate)	0.001	0.001	0.0018	0.001	0.003	0.001	0.005
Monozygotic twins	500	500	300	325	150	—	—
Sibs	35	40	35	20	20	100	8
First cousins	7	4	3	5	4	12	—
Second cousins	3	1.5	2	2	1.5	3	2

Data from Carter (1965).

gous twins, and this expectation requires no increase in similarity of any disease-related environmental factors in monozygous twins over dizygous twins. Table 1-4 records twin concordances for some well-studied diseases.

For nontwin relatives of individuals with a disease, the incidence of disease can also be obtained and compared with the population frequency. However, the age dependence in the prevalence of such traits as schizophrenia, adult onset diabetes, and hypertension has to be taken into account when incidence in relatives as, for example, between parents and children, is studied. Table 1-5 lists the proportions of affected family members of probands with various diseases relative to the incidence in the general population. For each disease there is a monotonic decrease in this ratio with the closeness of the relationship.

Twin concordance information has been combined with the familial incidence information to make an important inference in the case of diabetes mellitus. Table 1-6 shows that many more parents of concordant than of discordant monozygous twins were affected. This was especially so for those pairs in which the twin through which the disease was ascertained (the proband twin) was diagnosed after age 40. This difference suggests a heterogeneity in the disease (Cahill, 1979). Since the data of Table 1-6 were collected, association of juvenile onset diabetes with HLA, a set of genes defining the major histocompatibility complex, has been shown; this association is absent for mature onset diabetes.

Table 1-6. *Twin Concordance for Diabetes.* Number of concordant twin pairs among twin pairs with a diabetic parent.

	Concordant	Discordant
All ages	21 of 65 (32%)	1 of 31 (3%)
Proband twin diagnosed under age 40	6 of 30 (20%)	1 of 28 (3%)
Proband twin diagnosed after 40	15 of 35 (42%)	0 of 3

Tattersall and Pyke (1972), from Vogel and Motulsky (1979).

1.2 Genetic Variation in Populations

Genetic variation in populations may be partitioned into two categories depending on whether or not the variation produces an obvious phenotypic effect. Among the most important examples of such phenotypic variation in human populations are Mendelian genetic diseases. There is a slight ambiguity here in that some alleles of a given gene may cause disease while others may not. Some genes may therefore be considered in both classes. Genetic variation belongs to the qualitative category whether or not it results in obvious phenotypic variation.

Genetic variation can also be partitioned according to how it appears in the population. If the variation is apparent from a rather superficial examination of the population, then the population is said to be *polymorphic* for the examined trait and the relevant gene is said to show *allelic polymorphism*. If the trait of interest occurs sporadically in a very low frequency, the trait is said to be *ideomorphic* and the gene is said to have ideomorphic allelic variants. As an operational rule, a population is said to be polymorphic for a specific gene if the most common allele of that gene is less frequent than 99 percent.

1.2.1 Allelic Polymorphism

One of the most familiar allelic polymorphisms is eye coloration in *Homo sapiens* of European origin, where brown-blue difference is essentially due to a pair of Mendelian alleles. In natural populations of plants and animals obvious allelic polymorphisms are the exception rather than the rule. Nevertheless, there are several well-studied examples, some of which we now describe.

The European garden snail (*Cepaea nemoralis*) is conspicuously polymorphic for the color and pattern of the shell (Figure 1-18). Most populations of *Cepaea* are polymorphic, but the nature of the polymorphism varies among populations in a manner that appears to depend on the habitat (Figure 1-19). In *Cepaea nemoralis* shell color is determined at a single locus with alleles that produce, in decreasing order of dominance, dark brown, dark pink, pale pink, faint pink, dark yellow, and pale yellow. The presence or absence of bands on the shell is controlled by a single locus with two alleles, where the allele for unbanded shell pattern is dominant to that for banding. The banding

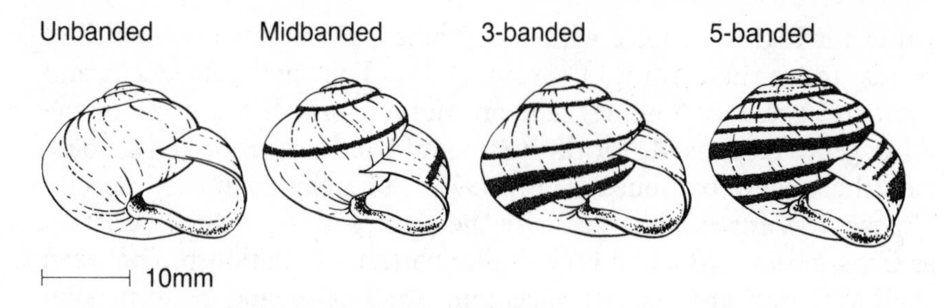

Figure 1-18. The banding pattern polymorphism of the garden snail, Cepaea nemoralis. *(After Jones et al., 1977, and André, 1973.)*

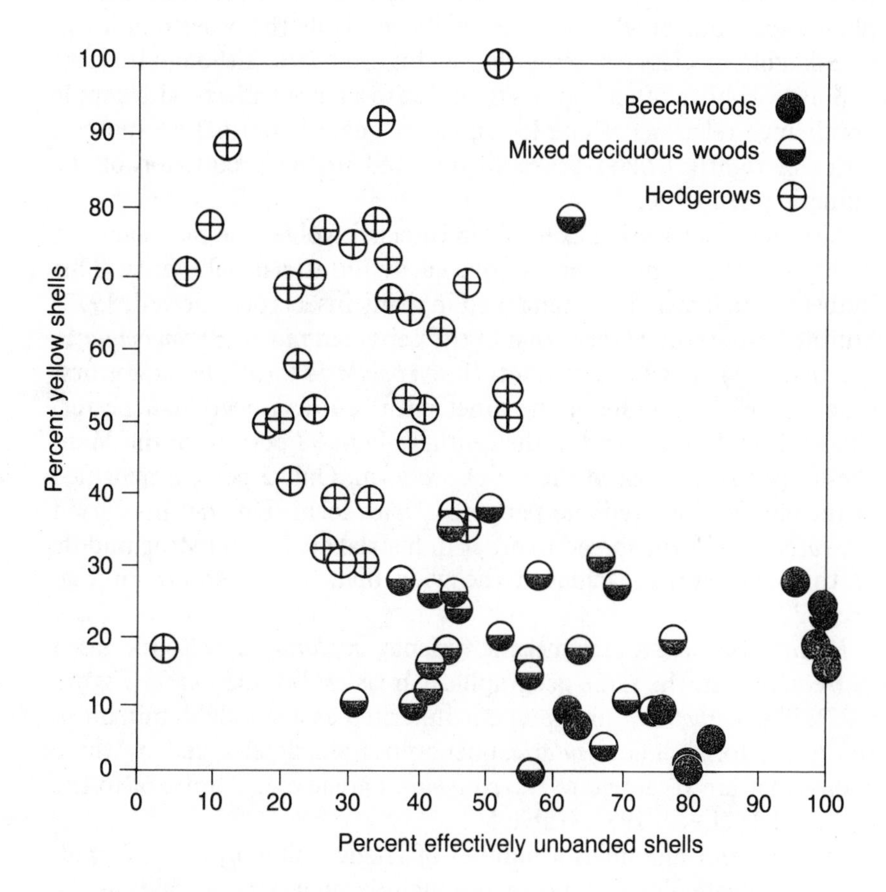

Figure 1-19. The difference among Cepaea nemoralis *populations in different habitats. (After Cain and Sheppard, 1954.)*

pattern is highly variable with the number, continuity, and width of bands under the control of several genes. The shell color locus and the band presence-absence gene are tightly linked, but some of the other genes responsible for the number of bands are unlinked to these (see the review by Jones et al., 1977). The closely related species *Cepaea hortensis* exhibits virtually the same pattern of shell variation as *Cepaea nemoralis* and a very similar pattern of relationship between shell variation and habitat variation. Shell color and pattern polymorphism is also found in many terrestrial and aquatic molluscs.

One of the most interesting allelic polymorphisms in insects involves Batesian mimicry. This phenomenon, most common in tropical butterflies, occurs when, in species palatable to a predator, there are two phenotypes, one of which is a normal form while the other mimics an unpalatable species coexisting with it (Figure 1-20). Although this sort of polymorphism often appears to be due to a pair of alleles with simple dominance relations at one locus, closer analysis normally exposes a series of tightly linked genes all involved in the production of the mimetic phenotype.

A second outstanding example in insects involves the prevalance of darkly colored forms in areas affected by industrial pollutants. This industrial melanism has been noted in many insects (Kettlewell, 1973) but the best-studied case is that of the peppered moth, *Biston betularia* (Figure 1-21). The black variant, *Biston betularia* f. *carbonaria* was first reported in the middle of the nineteenth century, near Manchester, England, and by the end of the century about 98 percent of the Manchester population was of the *carbonaria* form. Only 2 percent remained of the old light-colored and peppered *typica* form. During this period the *carbonaria* form spread to other industrial areas and by the middle of this century had achieved the wide distribution shown in Figure 1-22.

Meanwhile less extreme melanic variants, *insularia*, had also increased in frequency in the same geographical areas as did *carbonaria* (Figure 1-22). The *carbonaria* phenotype is inherited as a simple dominant to the *typica* form. The *insularia* phenotypes are determined by three alleles that are recessive to the *carbonaria* allele and dominant to the *typica* allele (Lees, 1981, Table 5.2).

In birds and mammals a number of allelic polymorphisms of plumage and coat color are known and include such familiar cases as the sex-linked polymorphism for yellow fur color of domestic cats. In the Arctic skua (*Stercorarius parasiticus*), for example, there is a plumage

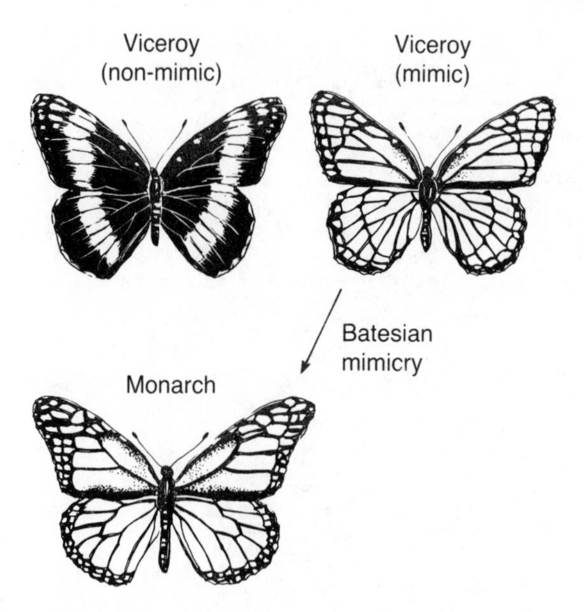

Figure 1-20. Batesian mimicry between the monarch and the viceroy butterflies. The monarch is unpalatable; part of the population of viceroy exhibits a wing pattern that mimics that of the monarch.

color polymorphism due to two alleles at one locus. The two homozygotes have light and dark plumage with the heterozygote intermediate (O'Donald, 1983).

Domesticated plants provide many examples of allelic polymorphisms including, of course, the garden pea used by Mendel in his experiments. In natural populations such simple variation is exceptional. There are, however, well-known allelic polymorphisms that govern the limitation of self-fertilization in flowering plants. The easiest of these to recognize involve flower morphology as, for example, in *Primula* (Figure 1-23). *Primula* populations are polymorphic with two common flower forms, called pin and thrum. Pin flowers have a long easily visible stigma with anthers placed deep inside the flower. Thrum flowers have a short stigma and the anthers are visible. At first glance, the pin flower looks female and the thrum male. In reality, both flowers are hermaphroditic, although a pin flower has a higher chance of insect pollination with pollen from a thrum flower and vice versa.

Figure 1-21. The cryptic effect of carbonaria *and* typica *morphs of* Biston betularia *in undisturbed (top) and polluted (bottom) habitats. (After Kettlewell, 1973.)*

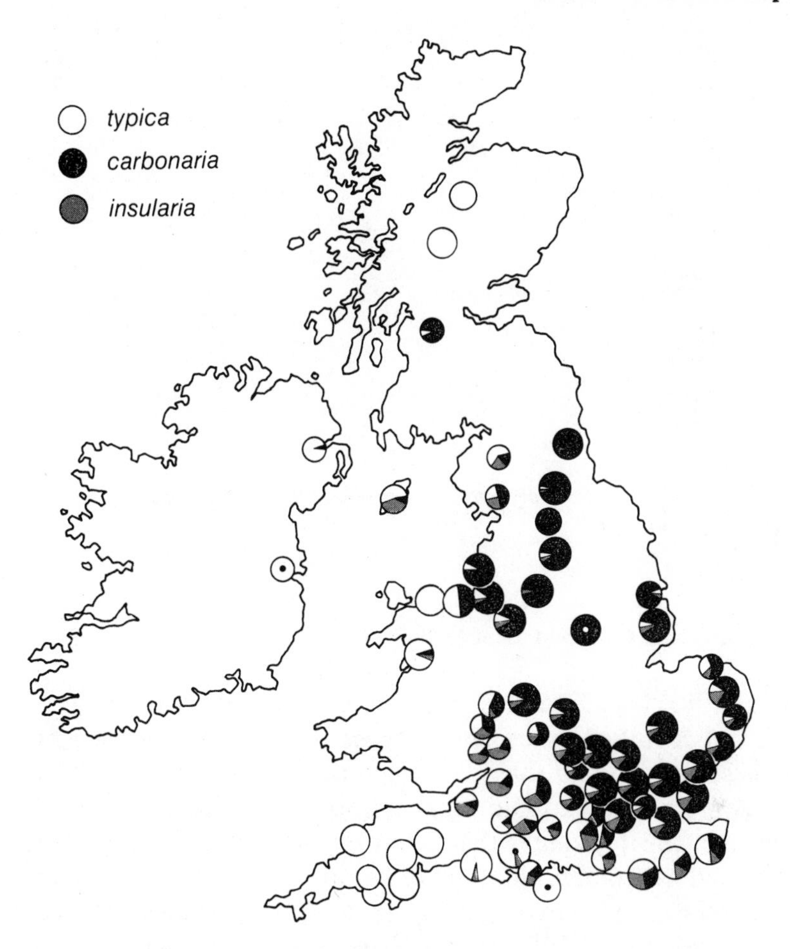

Figure 1-22. The distribution of the morphs of Biston betularia *in Britain. Pie size indicates sample size; sectors indicate the relative frequency of the three morphs. (After Kettlewell, 1973.)*

The flower morphology is inherited as if it were determined by one locus with two alleles, the thrums have the dominant phenotype and are usually heterozygotes, and the pins are recessive homozygotes. In fact, the polymorphism is governed by two tightly linked loci. One locus controls stigma length, with short dominant to long. The other controls another location, with thrum placement dominant to that of pin. The double dominant and double recessive are the thrum and pin phenotypes, respectively. Occasional crossovers probably account for the sporadic occurrence of plants with aberrant homostylous flowers.

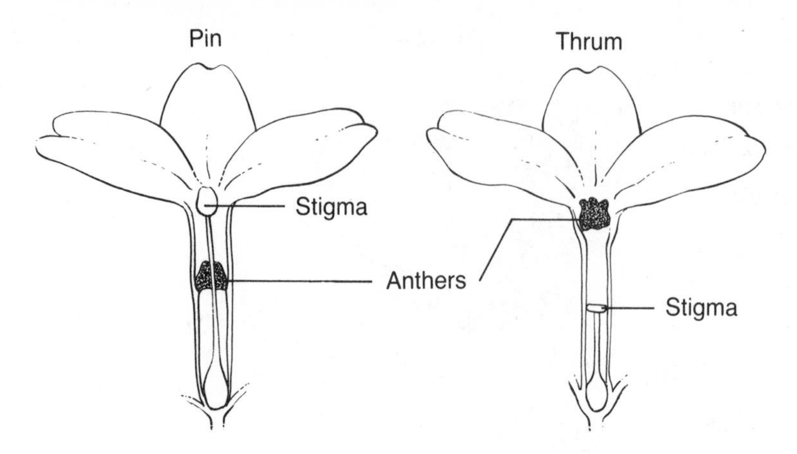

Figure 1-23. The pin-thrum polymorphism in primrose, Primula vulgaris. *(After Darwin, 1862.)*

1.2.2 Mendelian Genetic Diseases

McKusick (1983) lists 934 autosomal dominant, 588 autosomal recessive, and 115 X-linked human phenotypes thought to be caused by single genes, and most of these belong to the category of hereditary diseases. It is worth mentioning that the strict Mendelian definition of dominance, which requires that heterozygotes that carry one allele of the type which confers the trait in homozygotes should have the same phenotype as that homozygote, is not retained in human genetics. Rather, it is usual to call the disease dominant when it appears that heterozygotes exhibit a form of the disease or have a nontrivial potential to acquire it.

The majority of genetic diseases are caused by ideomorphic alleles. The total population incidence of confirmed single-gene diseases as outlined in Table 1-7 is approximately 1–2 per 1000 births with the three types of inheritance represented about equally often. Rather few diseases contribute a sizable fraction of this total incidence. Thus, among the dominant diseases, 16 percent of the population incidence is due to chondrodystrophy. Among the recessive diseases, cystic fibrosis, phenylketonuria (Figure 1-24), and albinism comprise 27 percent, 7 percent, and 5 percent, respectively, and among X-linked diseases hemophelia A and B comprise 20 percent and 6 percent and colorblindness 12 percent. The incidence of specific diseases, however,

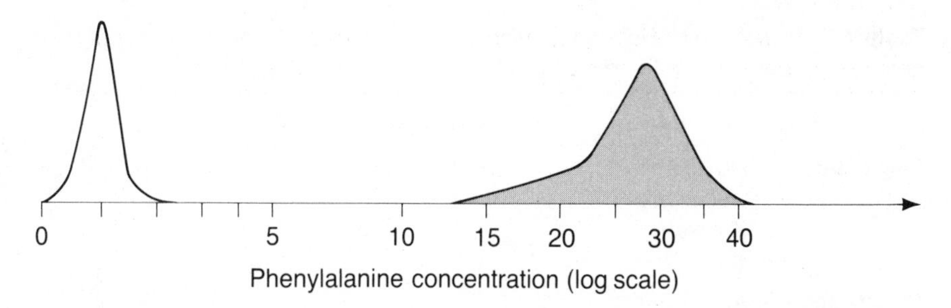

Figure 1-24. The variation in the concentration of the amino acid phenylalanine in blood plasma in normal children (left) and children with phenylketonuria (right). Patients with phenylketonuria lack the enzyme phenylalanine hydroxylase, so that the normal metabolism of phenylalanine is blocked. This results in elevated levels of phenylalanine in the brain and an abnormally high activity in alternative metabolic pathways. The result is mental retardation due to impaired development of the brain. (After Penrose, 1952.)

varies considerably from population to population. For instance, cystic fibrosis occurs in about one of every 2500 births in Caucasoid populations, whereas the incidence in Mongoloid and Negroid populations is one in every 250,000 births. In Ashkenazi Jews the frequency of Tay-Sachs disease is about 2×10^{-4}, while in Sephardic Jews and U.S. non-Jews the frequency is less than 1/100th of this. Although these are extreme examples, they illustrate the range of possible variations among populations.

Many of the Mendelian diseases are very serious, with drastic effects early in life if left untreated. In this group are many, mostly recessive metabolic diseases such as phenylketonuria, where the diseased individual lacks an important enzyme (see Suzuki et al., 1981, Table 10-3). Though other diseases may be equally damaging, the onset of the disease is delayed until later in life; an example is the dominant disease Huntington's chorea (Figure 1-25). This late onset of some diseases makes their genetic study difficult, because an individual may have carried a disease allele and passed it on to the offspring, but may have died before his onset of the disease. A dominant disease allele that has been carried by a normal individual is said to be *nonpenetrant*. This phenomenon of limited penetrance is well known for a number of dominant diseases.

A small class of hereditary diseases caused by single genes occurs at polymorphic frequencies in some populations. We describe three of

Table 1-7. *Incidence of Hereditary Disorders.* Incidence of different types of hereditary diseases in a large study of humans from British Columbia.

Aetiology		Number of cases per 10,000 liveborn	
Single gene:	Dominant	6	
	Recessive	9	
	X-linked	3	
	Total		18
Chromosome:	Autosomal	15	
	Sex	1	
	Total		16
Congenital malfunctions			358
Other multifactorial			158
Unknown			60
Total			610

From Trimble and Doughty (1974).

the best known: variants of the hemoglobin β chain, β-thalassemia, and G6PD deficiency. There are three common variants of the normal hemoglobin β allele *HbA* all due to single amino acid changes in the β-globulin chain. The variant alleles *HbS*, *HbC*, and *HbE* all cause anemia when homozygous, but only the sickle cell homozygotes, with genotype *HbS/HbS*, have an extremely severe form of the disease. Sickle cell anemia occurs in quite high frequencies in Central and West Africa. In Central Africa it can reach about 3 percent, and up to 30 percent of the population is heterozygous, *HbA/HbS*, showing the sickle cell trait. In West Africa the frequency of sickle cell trait is around 20 percent and in U.S. blacks it is about 10 percent. Hemoglobin C occurs mainly in West Africa in frequencies somewhat lower than those of hemoglobin S, and hemoglobin E is concentrated in Southeast Asia where it has been reported to reach a frequency of 30 percent in Cambodia and Thailand.

In the sickle cell polymorphism the three genotypes *HbA/HbA*, *HbA/HbS*, and *HbS/HbS* can be recognized phenotypically as normal, sickle cell trait, and sickle cell anemia, respectively. Thus we can assess the *genotypic* composition of the population at this locus as it coincides with the phenotypic composition. For the allelic polymorphisms considered in section 1.2.1, some of the phenotypic classes are genetically heterogeneous due to dominance. Knowing the genotypic composition of the population, we can describe the genetic composition of the population. Suppose that the numbers of three genotypes

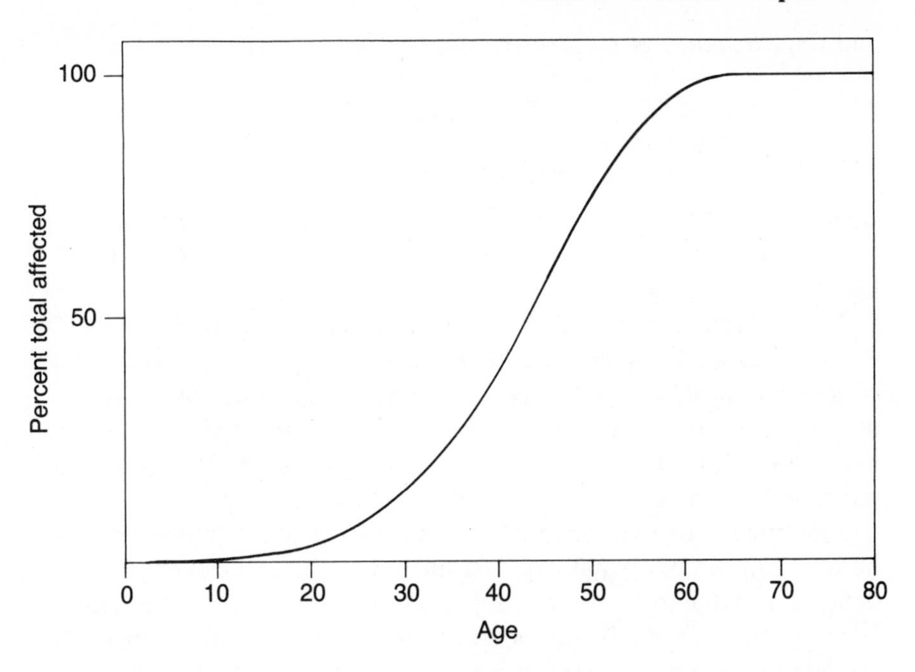

Figure 1-25. Distribution of onset age for Huntington's chorea among individuals having the disease gene, shown as the cumulative frequency of individuals in whom the disease is manifested before a given age. (After Wendt and Drohm, 1972.)

of the sickle cell polymorphism in the population are G_{AA}, G_{AS}, and G_{SS}, respectively. Homozygotes *HbA*/*HbA* each carry two *HbA* alleles and heterozygotes *HbA*/*HbS* each carry one *HbA* allele, so the total number of *HbA* alleles is

$$P_A = 2G_{AA} + G_{AS}.$$

Similarly, the total number of *HbS* alleles is

$$P_S = G_{AS} + 2G_{SS},$$

and the total number of alleles is $2N$ where $N = G_{AA} + G_{AS} + G_{SS}$ is the total number of individuals. The allele frequency of *HbA* is then

$$p_A = \frac{P_A}{2N} = \frac{2G_{AA} + G_{AS}}{2N}$$

and the frequency of *HbS* in the population is

$$p_S = \frac{P_S}{2N} = \frac{G_{AS} + 2G_{SS}}{2N}$$

with $p_A + p_S = 1$. In terms of allele frequencies, Central African populations can reach $p_S = 0.16$, West African $p_S = 0.10$, and U.S. black populations about $p_S = 0.05$.

A recessive disease which is almost always fatal is β-thalassemia. It is due to an allele called *HbT* that reduces or eliminates production of hemoglobin β chains. The *HbT* allele appears to alter the regulation of transcription and processing of mRNA for the synthesis of the hemoglobin β chain. The allele reaches frequencies of 10–20 percent in some African and a number of Mediterranean populations.

A deficiency of the enzyme glucose-6-phosphate dehydrogenase is quite common among males of African and Mediterranean origin. The deficiency is due to an X-linked gene, and in Figure 1-26 the activity levels of the enzyme in males and females are shown. Whereas the bimodality of the distribution in males allows an easy classification, the different female genotypes are not easy to infer from activity levels. The enzyme is encoded by an X-linked gene with two common alleles (A and B) that both produce normal enzyme activity. Allele A is found almost exclusively in populations of black African ancestry, whereas B is found in most other populations. In addition there are two common activity-deficient alleles causing a recessive disease. One allele, A^-, occurring in blacks, reduces the activity of the enzyme to 10–20 percent of normal, and the other, B^-, which is quite common in Mediterranean populations, causes an even more severe reduction of enzyme activity. Homozygotes (in females) and hemizygotes (males) for an activity-deficient allele show serious hemolytic reactions to a number of drugs. The disease has been studied intensively in Jews of North African origin, and in Kurdish Jews, for example, the frequency of affected males is over 50 percent.

These three genetic diseases occur in polymorphic frequencies only among populations that are heavily affected by the malarial parasite, *Plasmodium falciparum;* they are almost totally absent from other populations. The explanation for this dichotomy is believed to be that the disease-causing alleles mediate some resistance against the parasite in the heterozygotes (see section 3.2.2). Biochemical assays have been developed for all of these genes so that the presence of the disease allele can be detected in heterozygotes.

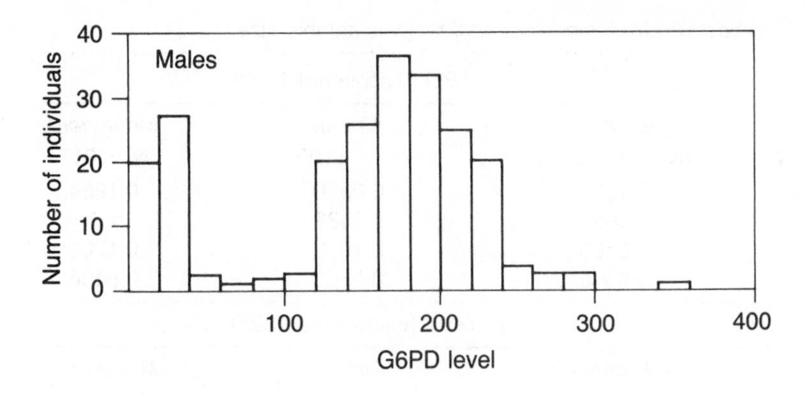

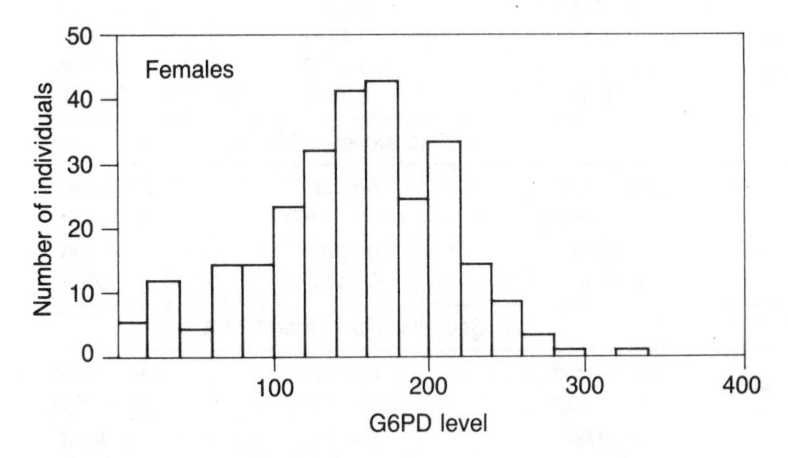

Figure 1-26. Distribution of red cell glucose-6-phosphate dehydrogenase activities in Nigeria. (After Harris, 1970.)

1.2.3 Blood Group Polymorphisms

The earliest human genetic polymorphism discovered was the *ABO* blood group system, and since the turn of the century a number of red cell surface antigen (i.e., blood groups) polymorphisms have been discovered. The blood groups are known for their importance in blood transfusions, and their discovery is linked to observed phenotypic variation in and among human populations. Accordingly, the most polymorphic blood group genes were discovered first, and only by increasingly extensive investigations were almost monomorphic blood group genes discovered (see Table 1-8).

Table 1-8. *Blood Group Allele Frequencies.* (*N* is sample size.)

	1. *ABO* (discovered 1900)		
	Caucasoid (N = 100,000)	*African* (N = 1109)	*Mongoloid* (N = 817)
A_1	0.1977	0.0970	0.1864
A_1	0.0683	0.0432	rare
B	0.0795	0.1078	0.1700
O	0.6545	0.7520	0.6436
	2. *MNS* (discovered 1927)		
	Caucasoid (N = 4116)	*African* (N = 292)	*Mongoloid* (N = 103)
MS	0.2415	0.0917	0.0405
Ms	0.3217	0.4481	0.5663
NS	0.0823	0.0788	0.0144
Ns	0.3545	0.3814	0.3788
	3. *P* (discovered 1927)		
	Caucasoid (N = 10,000)	*African* (N = 1162)	*Mongoloid* (N = 1381)
P_1	0.5546	0.7883	0.1406
P_2	0.4454	0.2117	0.8594
	4. *Secretor* (discovered 1930)		
	Caucasoid (N = 2435)	*African* (N = 300)	*Mongoloid* (N = 555)
Se	0.5078	0.4967	0.4818
se	0.4922	0.5036	0.5182
	5. *Rhesus* (discovered 1940)		
	Caucasoid (N = 8297)	*African* (N = 644)	*Mongoloid* (N = 4648)
CdE	0	0	0.0036
cDe	0.0186	0.7582	0.0334
CDe	0.4036	0.0376	0.7298
cDE	0.1670	0.0427	0.1870
CDE	0.0008	0	0.0041
CwDe	0.0198	0.0031	0
cde	0.3820	0.0997	0.0232
Cde	0.0049	0.0587	0.0189
cdE	0.0030	0	0
$C^w de$	0.0003	0	0

Based on data in Mourant et al. (1976) and Race and Sanger (1968), after Cavalli-Sforza and Bodmer (1971) and table 25 of Lewontin (1974).

Table 1-8. *Continued*

	6. *Lutheran* (discovered 1945)		
	Caucasoid (*N* = 2582)	*African* (*N* = 233)	*Mongoloid* (*N* = 36)
Lua	0.0415	0.0239	0.0
Lub	0.9585	0.9761	1.0

	7. *Kell* (discovered 1946)		
	Caucasoid (*N* = 10,156)	*African* (*N* = 1205)	*Mongoloid* (*N* = 1033)
K+	0.0419	0.0029	0.0
K−	0.9581	0.9971	1.0

	8. *Lewis* (discovered 1946)		
	Caucasoid (*N* = 1000)	*African* (*N* = 125)	*Mongoloid* (*N* = 85)
Le	0.8156	0.3188	0.7575
le	0.1844	0.6812	0.2425

	9. *Duffy* (discovered 1950)		
	Caucasoid (*N* = 1688)	*African* (*N* = 1162)	*Mongoloid* (*N* = 500)
Fy	0.0540	0.9944	0.0
Fy^a	0.4414	0.0026	1.0
Fy^b	0.5046	0.0030	

	10. *Kidd* (discovered 1951)		
	Caucasoid (*N* = 6546)	*African* (*N* = 176)	*Mongoloid* (*N* = 103)
JK^a	0.5100	0.7081	0.3103
JK^b	0.4900	0.2919	0.6897

	11. *Diego* (discovered 1955)		
	Caucasoid (*N* = 500)	*African* (*N* = 775)	*Mongoloid* (*N* = 300)
Di^a	0.0	0.0000	0.0305
Di^b	1.0	1.0000	0.9695

	12. *Xg* (sex-linked; discovered 1962)		
	Caucasoid (*N* = 2082)	*African* (*N* = 219)	*Mongoloid* (*N* = 549)
Xg^a	0.675	0.55	0.54
Xg	0.325	0.45	0.46

Table 1-8 lists the main alleles observed in Caucasoid, African, and Mongoloid populations at these blood group loci. Although there are rare alleles not recorded in the table, it should be noted that there are not a great many common alleles at any one locus.

Two blood groups, in particular the *ABO* system and the *Rh* system, have been the subject of much conjecture in the attempt to understand why they are polymorphic. One reason for this speculation is the occurrence of maternal-fetal incompatibility among certain mother-child genotype combinations. About 0.8 percent of the *A* or *B* offspring of an *O*-type mother, including the first born, develop hemolytic disease as a result of the passage across the placenta of anti-*A* or anti-*B* antibodies. Rhesus positive (*Dd*) offspring of Rhesus negative (*dd*) mothers, usually the second or later such births, are at risk for hemolytic disease, which can be more severe than that induced by ABO incompatibility. Both cases of incompatibility place heterozygous offspring of recessive homozygous mothers at risk. Further, *ABO*-incompatible mother-child combinations are less likely to suffer the consequences of *Rh*-incompatibility.

The *ABO* blood group has been shown to be significantly associated with a number of other noninfectious diseases. Duodenal and gastric ulcers are more likely to occur in group *O* patients than in others, while cancer of the stomach is more frequent in group *A* patients than in others. Vogel and Motulsky (1979, p. 166) provide a list of such significant associations. A sophisticated quantitative analysis of the effect of the association with duodenal ulcer on longer-term population frequencies of *A*, *B*, and *O* can be found in Cavalli-Sforza and Bodmer (1971). Their conclusion is that this effect is orders of magnitude smaller than that of *ABO* incompatibility.

Some blood groups show pronounced ethnic differences in allele frequencies. The frequency of the B allele is higher in Orientals than in Caucasoids (Table 1-8). The $R_o(cDe)$ chromosome is common in African blacks and almost absent elsewhere, while the Rhesus negative chromosome *cde* is more frequent in Caucasians than in other groups. The Di^a allele of the Diego system appears only in Orientals and American Indians, and provides evidence of the genetic association between these ethnic groups. The high frequency of the *Fy* allele of the Duffy system in Africans provides a major genetic differentiation from other groups. Using the blood group genes of Table 1-8 it is possible to develop measures of genetic distance between racial groups. Such measures are composites of the overall allele frequency differ-

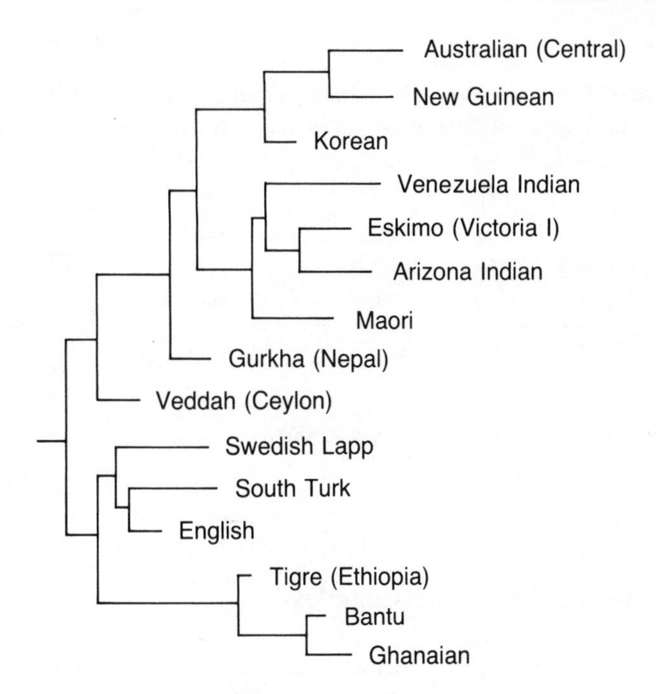

Figure 1-27. Evolutionary tree showing the genetic relationship among different human racial groups as calculated from allele frequencies at blood group loci. (After Cavalli-Sforza and Edwards, 1963.)

ences and can be used in combination with similar measures based on morphological variation to suggest the order in time of genetic divergence among the racial groups. One such picture, based on blood group allele frequencies, is drawn in Figure 1-27.

Exercise 1.2.A

In an investigation of the Rhesus blood types in the Danish population, the reactions to *C*-antiserum and *c*-antiserum were recorded. This reaction is determined by two codominant alleles *C* and *c*, and the phenotypic distribution was found to be

CC	*Cc*	*cc*	Total
1013	2764	1723	5500

From Gürtler and Henningsen (1954).

What are the allele frequencies in this population?

Exercise 1.2.B

Three investigations of *MN*-blood types in Greenland Eskimos produced the following phenotypic and genotypic distributions in the populations:

	MM	*MN*	*NN*	Total
Thule (north)	61	64	27	152
Julianehaab (south)	255	113	9	377
Angmagssalik (east)	475	89	5	569

From Fabricius-Hansen (1939, 1940); from Mourant et al. (1976).

a. Calculate the allele frequencies in these populations.

1.2.4 The Major Histocompatibility Complex

The immunological rejection of tissue transplanted from one human to another is due to genetically determined antigenic differences between the individuals. These antigens are called the *histocompatibility antigens*. The red cell-based *ABO* system is one such antigen system, but another, determined by antigens on the white blood cells and called the *HLA system* or the *major histocompatibility complex,* has received a great deal of research attention over the past two decades.

The HLA system comprises a group of four or five genes on chromosome 6 that span a map distance of about 1.8 cM, as depicted in Figure 1-28. The *A, B,* and *C* loci and their alleles are defined by reactions to specific antisera. There are two *D*-locus typing methods, one of which uses mixed lymphocyte culture (MLC) typing and produces an allelic series prefixed by *Dw*. A second method uses the reactions of specific antisera to cell suspensions enriched for B-lymphocytes. The resulting allelic series is prefixed by *DR*. No recombinants have been found between the two *D* series but it remains open as to whether they are in fact the same locus.

At first the main impetus for the development of research on HLA was its presumed importance in transplantation, but a recent focus has been the population genetic properties of the system. In Table 1-9, we see that the system is extremely polymorphic. In fact, the proportion of individuals who are heterozygous at HLA $-A$, $-B$,

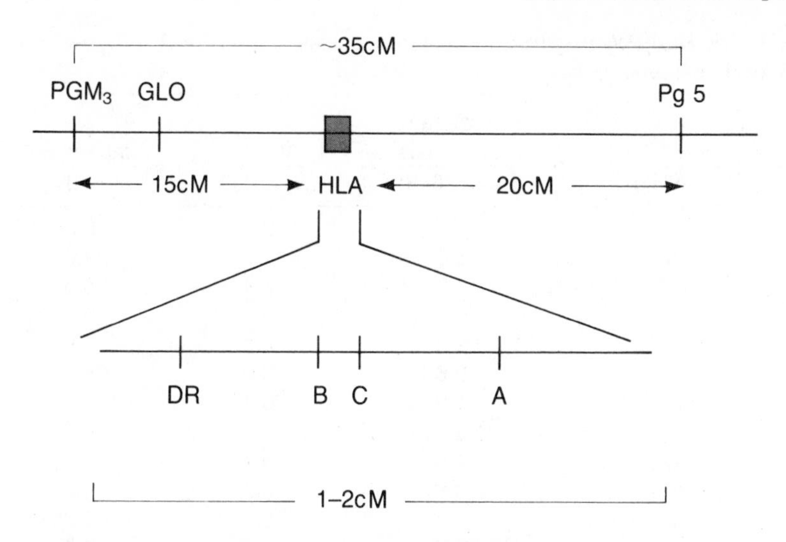

Figure 1-28. Simplified linkage map of the HLA region of the human chromosome 6 showing the A, B, C, *and* DR *loci. (After Vogel and Motulski, 1979.)*

and $-D$ is larger than 80 percent, and more than 75 percent of Caucasians, on average, will have four different *A* and *B* antigens. This is a far higher level of polymorphism than for any of the red cell blood groups. Unfortunately, there is no general explanation that would account for these high levels of polymorphism, but suggestions have been made that the variability in the HLA system has to do with responses to infectious agents. The analogue to the HLA system in the mouse is called the H-2 system and it is known to contain genes that govern immune responsiveness. It has been speculated that the extensive polymorphism in HLA may be the result of a continuing cycle of resistance development by the human host to pathogens that are themselves changing to overcome the resistance (see Bodmer and Bodmer, 1978).

Included in the general expression of polymorphism of the HLA system is the finding that certain alleles at pairs of loci are highly associated with one another. Thus the pairs (*A1, B8*) and (*Cw4, B35*) occur more often together in a chromosome than the separate allele frequencies would suggest. There are interesting differences among populations in these associations (Payne et al., 1977).

Table 1-9. *The HLA Polymorphism.* Allele frequencies at the HLA-*A*, −*B*, −*C* and *DR* loci in percent.

A Locus	Allele	European Caucasoids N = 2648	African N = 367	Japanese N = 949
	A1	14.9	3.3	0.5
	A2	26.0	14.7	24.6
	A3	11.6	7.4	0.5
	A11	5.9	0.6	9.0
	Aw23	2.3	10.8	0.5
	Aw24	9.6	2.9	35.6
	A25	1.9	0.4	0.1
	A26	3.7	3.8	9.8
	A28	4.0	8.7	0.5
	A29	3.8	6.3	0.2
	Aw30	2.4	15.4	0.2
	Aw31	2.7	2.2	8.0
	Aw32	4.5	1.5	0.1
	Aw33	1.7	4.6	6.8
	Aw34	0.6	6.5	1.0
	Aw36	0.3	1.7	0.3
	Aw43	0.0	1.0	0.0
	blank	4.3	8.4	2.5

As with the red cell polymorphisms, there are racial differences in allele frequencies. Bodmer and Bodmer (1978) delimit three classes of alleles: those like *A2* common in all populations, those like *A1, A3, A28, A29, Aw30,* and *Aw32,* which are rare or absent from one or more groups (the Japanese) but present in most groups, and those like *Aw43* and *Bw42,* which are specific to one group (African blacks).

Another focus of interest in the HLA system is summarized in Table 1-10, which illustrates the most important diseases in which patients have a highly increased frequency of a specific antigen over controls. This suggests some role for the HLA gene products in the etiology of these diseases, although no specific biochemical relationships have yet been detected. These disease associations can, when strong, aid in diagnosis, although it must be remembered that the disease is in fact being used as a marker for the HLA allele, not the reverse. Also, of the diseases in Table 1-10 only hemochromatosis appears to be a Mendelian disease. The others are of complex etiology, and although they show familial aggregation, the role of specific genes is yet to be determined.

Table 1-9. *Continued*

B Locus	Allele	European Caucasoids N = 2652	African N = 365	Japanese N = 950
	B7	8.8	8.9	5.9
	B8	8.2	2.9	0.1
	B13	2.8	0.7	2.0
	B14	3.0	4.1	0.1
	B18	5.8	3.9	0.0
	B27	3.9	1.5	0.4
	Bw35	9.5	6.2	7.3
	B37	1.5	0.4	0.5
	Bw38	2.5	0.0	0.2
	Bw39	2.1	1.8	2.9
	Bw41	1.0	1.2	0.4
	Bw42	0.3	7.7	0.6
	Bw44	11.0	7.1	6.5
	Bw45	1.1	3.9	0.2
	Bw47	0.4	0.1	0.2
	Bw48	0.5	1.1	2.3
	Bw49	2.3	2.5	0.3
	Bw50	1.2	0.7	0.0
	Bw51	7.2	1.4	8.3
	Bw52	1.5	1.0	10.9
	Bw53	0.9	6.5	0.1
	Bw54	0.0	0.0	7.3
	Bw55	2.2	0.8	2.9
	Bw56	0.6	0.0	1.1
	Bw57	3.1	3.9	0.0
	Bw58	1.1	10.7	0.9
	Bw59	0.5	0.8	2.1
	Bw60	3.4	1.4	6.6
	Bw61	1.7	0.4	8.8
	Bw62	5.3	1.0	8.8
	Bw63	0.5	0.3	0.2
	blank	6.1	17.1	12.1

Table 1-9 continued on page 50.

An important potential role for the HLA polymorphism is in classifying heterogeneous diseases. Thus *B8, Bw15,* and especially *DRw4* are increased in patients with juvenile onset diabetes, but there is apparently no such association with mature onset diabetes. This association also exemplifies the tenuous connection between such associations and the genetics of the disease, since juvenile onset diabetes shows substantially less familial aggregation than does mature onset diabetes.

Table 1-9. *Continued*

C Locus	Allele	European Caucasoids N = 2651	African N = 365	Japanese N = 950
	Cw1	4.1	0.1	17.6
	Cw2	5.1	12.0	0.4
	Cw3	10.1	9.2	26.9
	Cw4	12.1	15.9	4.7
	Cw5	6.0	2.9	0.1
	Cw6	7.9	9.0	0.7
	Cw7	2.3	2.4	1.1
	Cw8	1.9	0.4	0.1
	blank	50.6	47.7	48.5

DR Locus	Allele	European Caucasoids N = 2499	Negroes N = 323	Japanese N = 884
	DR1	6.9	4.9	6.3
	DR2	13.4	15.4	20.0
	DR3	10.8	17.3	1.6
	DR4	9.6	4.9	23.5
	DR5	10.3	13.3	2.2
	DRw6	2.2	5.3	4.6
	DR7	12.5	9.8	0.5
	DRw8	2.7	5.6	6.5
	DRw9	1.1	2.7	12.2
	DRw10	0.7	1.9	0.6
	blank	29.8	19.0	22.0

Data taken from the Eighth International Histocompatibility Workshop (Terasaki, ed., 1980). The symbol *w* indicates that the allelic assignment is not final.

1.2.5 Protein and Enzyme Polymorphism

By the mid-1950s protein electrophoresis had been used to reveal genetic variation in human populations for such genes as those responsible for the blood plasma proteins haptoglobin (a transport protein that can bind hemoglobin) and transferrin (Table 1-11). The alleles of these genes are defined by the electrophoretic mobility of the proteins they produce. Electrophoresis was also used in the diagnosis of heterozygotes for hemoglobin variants (see section 1.2.2).

The acid phosphatase example of Figure 1-4 is a case where the identification of separate genotypes (by electrophoresis) has enabled

Table 1-10. *Diseases with HLA Associations.*[a]

Disease	Antigen[b]	Frequency (%)		Relative risk[e]
		Patients	Controls	
Coeliac disease[c]	*DRw3*	96	27	64.5
	B8	67	20	8.1
Chronic active hepatitis[d]	DRw3	41	17	3.4
	B8	52	15	6.1
Myasthenia gravis[d]	DRw3	32	17	2.3
	B8	39	17	3.1
Graves' disease[c]	DRw3	53	18	5.1
	B8	44	18	3.6
Juvenile onset diabetes[d]	*DRw3*	27	17	1.8
	B8	32	16	2.5
	DRw4	39	15	3.6
Rheumatoid arthritis[d]	DRw4	56	15	7.2
Myasthenia gravis: Japanese[d]	*DRw4*	59	35	2.7
Juvenile onset diabetes: Japanese[d]	*DRw4*	65	35	3.4
Multiple sclerosis[d]	*DRw2*	41	22	2.5
Ankylosing spondylitis[c]	*B27*	90	8	103.5
Reiter's disease[c]	*B27*	80	9	40.4
Haemochromatosis[d]	*A3*	72	21	9.7
Psoriasis: Caucasoids[d]	*Cw6*	50	23	3.3
	B13	23	5	5.7
	B17	19	9	2.4
	B37	5	2	2.6
Psoriasis: Japanese[d]	*Cw6*	53	7	15
	B13	18	1	22
	B37	35	2	26
	A1	30	2	21

a. All individuals were Caucasoids, unless otherwise indicated.
b. Antigens originally designated Dw have been replaced by DRw here. The specificity *B17* has been replaced in the nomenclature of the Eighth International Histocompatibility Workshop (Terasaki, ed., 1980).
c. Data from Dausset and Svejgaard (1977).
d. Data from Seventh International Histocompatibility Workshop (Bodmer and Bodmer, 1978).
e. Relative risk estimate is given by $p_d(1 - p_c)/p_c(1 - p_d)$, where p_d is the frequency in patients and p_c is the frequency in controls.

the general population distribution of enzyme activities to be broken down into separate distributions for each genotype. In Figure 1-29 we see that each has its own distinct mode. It should be stressed that only

Table 1-11. *Plasma Protein Polymorphisms.*

		Haptoglobins[a]	
	Caucasoid (*N* = 13,262)	*Negroid* (*N* = 551)	*Mongoloid* (*N* = 172)
Hp^1	0.3871	0.5561	0.2820
Hp^2	0.6129	0.4439	0.7180
		Transferrin	
	Caucasoid (*N* = 2395)	Negroid (*N* = 357)	Mongoloid (*N* = 300)
Tf^C	0.9948	0.9706	0.9783
Tf^D	0.0006	0.0252	0.0217
Tf^{B1}	0.0008	—	—
Tf^{B2}	0.0038	0.0042	—

Data from Mourant et al. (1976).
a. Hp^1 can be subdivided into HP^{1S}, Hp^{1F}, the relative frequencies of which seem to vary widely among different European studies.

after the independent identification of the genotypes was such a decomposition possible. The reverse process of disassembling a phenotype distribution into separate genotypic components is a central component of much of modern population genetics.

The development of the methodology for assaying enzyme variation by electrophoresis in the 1950s and 1960s allowed the investigation of the variation at a large number of genes coding for these proteins by disclosing subtle biochemical variation that need not have any obvious phenotypic effect. Thus, genetic variation in the primary gene products could be assessed without referring to phenotypic variation.

The first studies at the population level of electrophoretically detectable variation in enzymes were published in 1966 (Harris, 1966; Lewontin and Hubby, 1966). Since that time a vast number of population surveys of plants and animals have uncovered a substantial amount of genetic polymorphism (Nevo et al., 1984). The number of loci (coding for these enzymes) that are polymorphic varies from species to species and the level of polymorphism measured either by number of alleles present, or proportion of the sample heterozygous, also varies substantially. Insects generally turn out to have the highest level of this sort of genetic polymorphism (Table 1-12), while humans and higher vertebrates are among the least polymorphic at these enzyme

Table 1-12. *Enzyme Polymorphism in Fruitflies.* Reviews data on electrophoretic variation in *Drosophila willistoni* of Ayala et al. (1974). For each locus the total number of observed alleles is given as well as indications of the degree of polymorphism. Polymorphism at the 5% level (+) means that the most common genotype has a frequency of 95% at most; the 1% level is similarly defined. The last column shows *H*, the frequency of heterozygotes at the various loci.

Locus	Number of alleles	Polymorphism[a]		H
		5%	1%	
Lap-5	7	+		.564
Est-2	7	−	+	.082
Est-3	3	−	+	.067
Est-4	2		−	.000
Est-5	3	−	+	.088
Est-7	10	+		.667
Acp-1	5	−	+	.049
Ald-1	4	+		.166
Ald-2	5	+		.178
Adh	4		−	.012
Mdh-2	3		−	.007
αGpdh	3		−	.010
Fh	4	+		.090
Got	6	−	+	.038
G6pdh	8	+		.197
Hbdh	4		−	.015
Icd	4	−	+	.022
Odh-1	7	+		.165
Me-1	3	−	+	.049
Me-2	7	+		.281
Xdh	10	+		.670
Ao-1	11	+		.554
Ao-2	9	+		.472
Sod	3		−	.019
Tpi-2	3	−	+	.025
Pgm-1	5	+		.137
Ak-1	7	+		.592
Ak-2	6	−	+	.038
Hk-1	5	−	+	.064
Hk-2	6	+		.165
Hk-3	2		−	.000
Average	$\frac{166}{31} = 5.4$			.177

a. 14/31 = .45 is the fraction of genes polymorphic by the 5% criterion, and 24/31 = .77 are polymorphic by the 1% criterion.

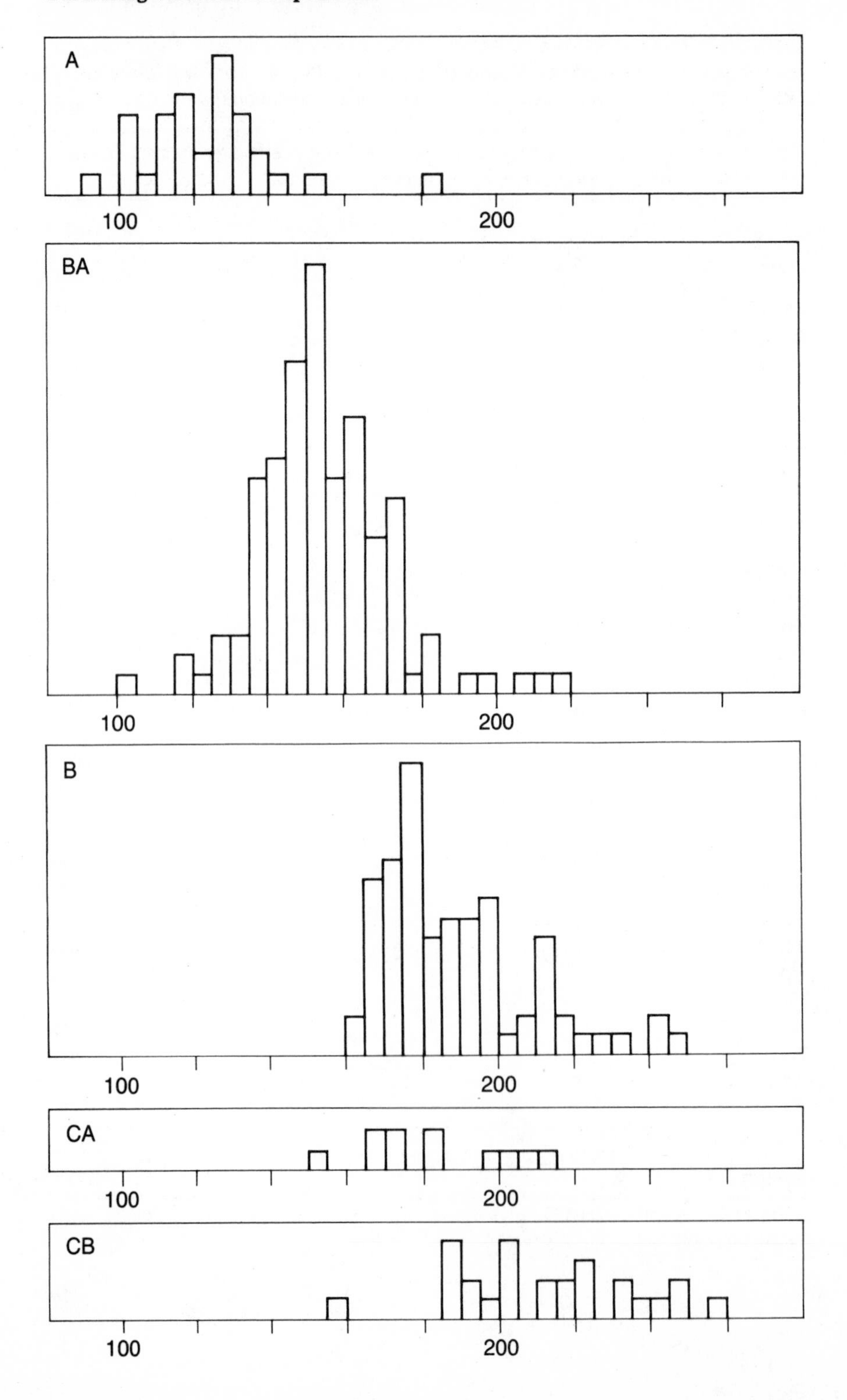
A
100
200
BA
100
200
B
100
200
CA
100
200
CB
100
200

loci. Nevertheless, of the 71 loci tested in humans up to 1972, 20 are polymorphic and the fraction of the population heterozygous at one of these polymorphic loci is 6–7 percent. Some of this variation is exhibited in Table 1-13.

From the genetic code it can be estimated that about one-third of base substitutions produce amino acid replacements which cause a change in the charge of a protein and therefore a change in its electrophoretic mobility. Thus we may be detecting only one-third of all the amino acid substitutions that exist, and the level of polymorphism at these enzyme loci may be even greater than our current observations. However, more sensitive techniques of electrophoresis have been developed and used in recent years (see, for example, Singh et al., 1976) and there are recent suggestions, based on studies of known amino acid changes in hemoglobin, that these more sensitive methods reveal up to 85 percent of existing amino acid substitutions (Ramshaw et al., 1979).

Exercise 1.2.C

In Table 1-13 consider the gene frequencies for pseudocholinesterase in African and Mongoloid populations. Are they different?

Exercise 1.2.D

In an investigation of the red cell acid phosphatase polymorphism in a German population, Hummel et al. (1970) found the following electrophoretic phenotypes

A	BA	B	CA	CB	C	Total
199	707	663	80	149	2	1800

where A, B, and C correspond to the presence of specific electrophoretic bands.

Calculate the allele frequencies at the acid phosphatase locus under the assumption that the alleles are codominant.

Figure 1-29 (opposite). Distribution of red cell acid phosphatase activities in individuals of different electrophoretic phenotypes (cf. Figure 1-4). Histograms show the number of individuals having activities in the various intervals; totals from the top are 33, 124, 81, 11, and 26, respectively. (After Spencer, Hopkinson, and Harris, 1964.)

Table 1-13. *Enzyme Polymorphism in Humans.*

	1. *Pseudocholinesterase*		
	Caucasoid (*N* = 952)	*African* (*N* = 102)	*Mongoloid* (*N* = 120)
E_1^u	0.9984	0.9951	1.0
E_1^o	0.016	0.0049	0.0
	2. *Acid phosphatase*		
	Caucasoid (*N* = 1800)	*African* (*N* = 577)	*Mongoloid* (*N* = 620)
p^a	0.3292	0.1898	0.2218
p^b	0.6061	0.8041	0.7782
p^c	0.0647	0.0061	0.0
	3. *G6PDH*		
	Caucasoid (*N* = 404)	*African* (*N* = 1451)	*Mongoloid* (*N* = 1177)
GdA^+		0.2295	
GdB^+	1.0000[a]	0.5541	0.9626
GdA^-		0.2164	
GdB^-			0.0374
	4. *6PGDH*		
	Caucasoid (*N* = 4558)	*African* (*N* = 693)	*Mongoloid* (*N* = 228)
PGD^A	0.9779	0.9394	0.9342
PGD^C	0.0214	0.0592	0.0658
	5. *Adenylate kinase*		
	Caucasoid (*N* = 2008)	*African* (*N* = 688)	*Mongoloid* (*N* = 227)
Ak^1	0.9626	1.0	0.9980
Ak^2	0.0359	0.0	0.0020
Ak^3	0.0015	0.0	0.0
	6. *Phosphoglucomutase*		
	Caucasoid (*N* = 2115)	*African* (*N* = 593)	*Mongoloid* (*N* = 276)
PGM^1	0.7641	0.8339	0.7482
PGM^2	0.2348	0.1661	0.2500

Data from Mourant et al. (1976).
a. Mediterraneans have 1 − 40% of B^-.
b. Philippine.

Table 1-13. *Continued*

		7. Adenosine deaminase	
	Caucasoid (N = 861)	*African* (N = 213)	*Mongoloid* (N = 100)[b]
ADA_1	0.9355	0.9883	0.8850
ADA_2	0.0645	0.0117	0.1150

1.2.6 Restriction Fragment Length Polymorphism

Recombinant DNA technology has made it possible to go beyond the serologic and electrophoretic methods of detecting genetic variation. We may now detect and study the properties of variation in coding and noncoding sections of the DNA itself. The use of restriction enzymes has enabled the revelation of DNA polymorphism at the level of one or a few nucleotides. Each restriction enzyme recognizes a specific code, the recognition site, that is only a few bases long and shears the DNA at that point. Upon digestion of the DNA with the enzyme, pieces whose lengths are determined by the presence or absence of recognition sites are made (Fig. 1-30). By gel electrophoresis these may be separated by size, and a radioactive (or otherwise marked) DNA probe is used to visualize the size categories containing a specific gene. For example, in a restriction analysis of the gene cluster producing the β-globin of human hemoglobin such a probe would be made by the action of the enzyme reverse transcriptase on purified β-globin mRNA, which produces DNA complementary to the RNA (called cDNA). The radioactive probe hybridizes with its complementary DNA, after which it can be visualized by autoradiography.

The technique just described allows the detection of the presence or absence of restriction enzyme recognition sites by the variation in the segment length of the DNA pieces hybridizing to specific probes. Polymorphisms detected in this way are called *restriction fragment length polymorphisms* (*RFLP*) and, so far, more than 200 such polymorphisms have been detected in the human genome. Figure 1-30 shows how this works in the case of the β-globin region on chromosome 11, where more than 25 RFLPs are known. While most of these polymorphisms occur across all racial groups examined, some are restricted to one or few races. To date, RFLPs have been detected on practically all chromosomes in humans (O'Brien, 1984).

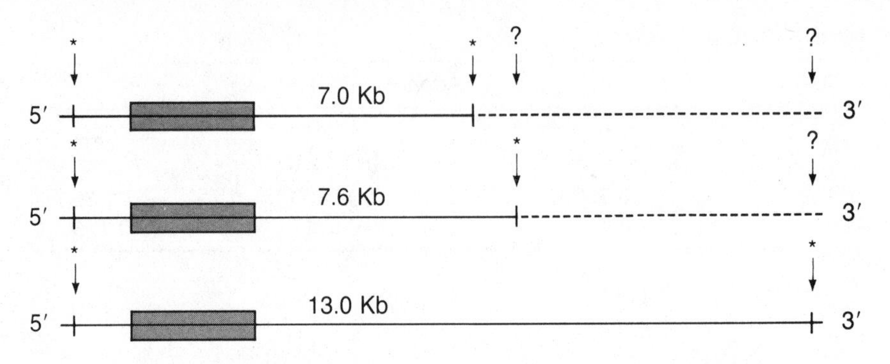

Figure 1-30. Variation in length of restriction fragments containing the β-globulin gene (heavy sector) as revealed in DNA digests using the restriction enzyme HpaI from the virus Hemophilus parainfluenzae. *This enzyme recognizes the base sequence 5′-*GTTAAC*-3′, and this recognition site occurs close to the β-globulin gene and 5′ to it (the site marked * to the left). Occurrence of the recognition site in the 3′ direction from the β-globulin gene varies in the populations studied and may produce a 7.0 kb, a 7.6 kb or a 13.0 kb (kilobase = 1000 bases) DNA fragment containing the gene. Restriction sites producing these DNA fragments in the* HpaI *digest are marked by *. The recognition site might occur at the locations marked ?, but the fragments so produced cannot be revealed by the β-globulin probe. The* HbS *allele is strongly associated with the 13 kb fragment, which lacks the recognition sites producing the 7.0 kb and 7.6 kb fragments.*

In *Drosophila melanogaster,* Langley and colleagues (1982) found RFLPs in a 12,000-nucleotide region containing the alcohol dehydrogenase gene, *Adh*. In most populations *Adh* exhibits two electrophoretic alleles, one fast-migrating Adh^f and one slow-migrating Adh^s form. The difference between these is due to a change in one amino acid. Recently, Kreitman (1983) has taken the search for variation one stage further, to the level of the *Adh* DNA sequence itself. In 11 *Adh* genes from five populations he found 43 nucleotide sites to be polymorphic, 14 of these in the *Adh* coding region. Only one of the 14 produced an amino acid change (the change from Adh^f to Adh^s).

On the basis of restriction enzyme recognition site variation, current estimates are that about one in every few hundred nucleotide sites should show polymorphic variation. These figures should be viewed circumspectly because restriction sites may not represent random DNA sequences, and there is a statistical difference between the frequency of those causing amino acid changes and those that do not.

The use of DNA polymorphisms as genetic markers in prenatal diagnosis of single-gene diseases was pioneered by Kan and Dozy (1978). They showed that the sickle cell hemoglobin gene, *HbS* (section 1.2.2), in blacks of West African origin is associated with a 13 kb DNA fragment not usually seen in Caucasians or healthy black individuals. Kan and Dozy suggested that when the parents carry the 13.0 kb DNA fragment, a restriction enzyme analysis of DNA from amniotic fluid could be useful in prenatal diagnosis of sickle cell anemia. Of course, such an approach is predicated on tight linkage between the DNA marker and the Mendelian disease gene. This method of antenatal diagnosis has been used successfully for the major hemoglobinopathies sickle cell anemia and thalassemia, and in fact has been replaced by more direct methods of DNA analysis. These methods are being extended to other biochemical abnormalities such as phenylketonuria, where the cells in the aminotic fluid may not express the abnormal phenotype.

2

Population Distribution of Alleles and Genotypes

Genetic differences among individuals are transmitted according to the laws of genetics. These laws allow us to predict the genotypes of offspring of parents with known genotypes and to relate the genotypic variation among a population of offspring to that in the parental population.

Since the laws of genetics are probabilistic, their extension to laws of population genetics are also probabilistic. This chapter will derive the basic rules of population genetics as simple deductions from the basic laws of genetics: the laws of Mendelian segregation and Morgan's rules of recombination.

2.1 Genetic Transmission in Populations

Mendelian genetics describes the way in which hereditary information is passed from parents to offspring by genes. The basic laws of population genetics are concerned with the description of the genes in a population as a function of the genes in the parents of that population. The simplest, almost trivial rule is that each of the genes in the offspring population is a copy of one of the genes carried by the parents. In this chapter the possibility of mutational change will be ignored and deferred until Chapter 3. We proceed to show how the simple laws of Mendelian segregation allow a surprisingly detailed account of the distribution of alleles in the offspring population.

2.1.1 Allele Frequency Conservation for Autosomal Genes

The *law of Mendelian segregation* states that the two genes at an autosomal locus are represented equally often in the gametes produced by an individual. This formulation of Mendel's law has an immediate generalization applicable to the transmission of the genetic material between generations at the level of the population. Consider the collection of mature adult individuals about to reproduce; we refer to this population as the *breeding population*. Each of these individuals produces gametes in such a way that the copies of the two genes at an autosomal locus are equally frequent. Thus if all males, say, produce equally many gametes, then copies of the genes carried by the breeding males are represented equally often among their gametes. This is Mendel's law of segregation on the level of the population: *If each breeding individual produces the same number of gametes, then the genes in the breeding population are represented equally often among the gametes produced.*

This population law of segregation can be extended to a law of heredity at the population level. If each breeding individual produces an equal number of offspring, then it also contributes the same number of *gametes* to the formation of the offspring population. Thus each gene among the breeding individuals is represented equally often among the gametes forming the offspring population, and therefore among the genes carried by the offspring. This argument neglects the existence of two sexes, but the basic idea is correct. If all females produce an equal number of offspring, and if all males fertilize an equal number of the eggs to form the offspring, then each gene among the breeding females is represented equally often among the egg cells that produce the offspring, and among the breeding males each gene is equally frequent among the fertilizing sperm. The assumption of Mendelian segregation therefore allows prediction of the genes in the offspring population from the genes in the breeding population under the assumption that each breeding individual produces the same number of offspring.

Consider an autosomal locus with the alleles A_1 and A_2, and suppose that the three genotypes A_1A_1, A_1A_2, and A_2A_2 occur in the numbers G_{11}, G_{12}, and G_{22} among breeding females, with $N = G_{11} + G_{12} + G_{22}$ the total number of breeding females. Then the frequency of A_1 alleles among the breeding females is then

$$p_1 = \frac{2G_{11} + G_{12}}{2N},$$

and the allele frequency of A_2 is

$$p_2 = \frac{G_{12} + 2G_{22}}{2N},$$

where $p_1 + p_2 = 1$. From the segregation law, the frequency of A_1 and A_2 in the egg cells will be p_1 and p_2, if all females produce equally many eggs. This restrictive condition is not necessary. We need only assume that the females of each genotype, on average, produce equally many eggs. Thus if the female genotypes on the average produce equally many offspring, then the frequencies of A_1 and A_2 among the egg cells giving rise to these offspring are expected to be p_1 and p_2, the allele frequencies in the breeding females. Similarly, if the male genotypes on average fertilize equally many of these eggs, then the frequencies of A_1 and A_2 among the fertilizing sperm are expected to be q_1 and q_2, where

$$q_1 = \frac{2H_{11} + H_{12}}{2M},$$

$$q_2 = \frac{H_{12} + 2H_{22}}{2M},$$

and H_{11}, H_{12}, and H_{22} are the numbers of the genotypes A_1A_1, A_1A_2, and A_2A_2 among the breeding males, with $M = H_{11} + H_{12} + H_{22}$. Again we have $q_1 + q_2 = 1$. Since each offspring is formed by one male and one female gamete, the expected frequency of A_1 among the offspring is $(p_1 + q_1)/2$ and the expected frequency of A_2 is $(p_2 + q_2)/2$. In other words, the allele frequencies among the offspring zygotes equal those in the breeding population, where allele frequency is understood as the average of the allele frequencies in breeding males and breeding females. In the simplest case, where the allele frequencies in the two sexes are equal, $p_1 = q_1$ and $p_2 = q_2$, we expect these frequencies among the offspring to equal those in the breeding populations of either sex. From the independent segregation of chromosomes, the two alleles at a given autosomal locus in any male are represented equally often among his X- and Y-carrying sperm. Thus, the allele frequencies among the X- and Y-carrying sperm forming the zygotes will be equal, so the allele frequencies among male and female zygotes will be the same. The law of heredity for an autosomal locus at the population level is then: *If breeding individuals of each genotype on average produce equally many offspring, and if segregation is Mendelian, then the allele frequencies among the offspring zygotes of either sex equal the allele frequencies in the breeding population* (Table 2-1).

Exercise 2.1.A

Consider the data in Table 2-1.

a. Explain why the classes A_1A_1 mother – A_2A_2 offspring and A_2A_2 mother – A_1A_1 offspring are not observed.
b. Find the allele frequencies q_1 for A_1 and q_2 for A_2 among the males that fertilized the A_1A_1 females.
c. Same question for A_2A_2 females.
d. What frequency of A_1A_2 offspring is expected from A_1A_2 mothers?
e. What are the allele frequencies among the males that fertilized the A_1A_2 females? (Check the answer to the previous question and satisfy yourself that it shows that the class A_1A_2 mother – A_1A_2 offspring is uninformative about the fathers.)
f. Count the number of observed A_1 alleles in paternal gametes you can be sure about and similarly for A_2 alleles.
g. Find the allele frequencies among the breeding males.
h. Find the allele frequencies among the mothers.
i. What are the allele frequencies among zygotes?
j. Why are the answers in question (i) slightly different from the answers predicted from (g) and (h)? (This is not an easy question, but think it over anyway! The answer is *not* selection.)

Phenomena that cause deviations from this law of heredity at the population level is referred to as *natural selection*. Violation of the assumption of equal number of offspring for individuals of the different genotypes is referred to as *fecundity selection*, while deviation from the assumption of Mendelian segregation is called *gametic selection*. The latter can be the result of meiotic drive or differential gametic viability. Selection may occur at other stages of the life. Differences in the survival ability of individuals of different genotypes, from their formation as zygotes until adulthood, are referred to as *zygotic selection* or *differential zygotic viability*, while differences in the probability of breeding among adult individuals of various genotypes are called *sexual selection*.

Table 2-1. *Mother-offspring Combinations.* Data on an esterase polymorphism in *Zoarces viviparus.* The mothers are caught while pregnant, and the genotype of the mother as well as the genotype of one randomly selected offspring is determined. See Exercise 2.1.A.

Genotype of mother	Genotype of offspring			Genotypic distribution among mothers
	A_1A_1	A_1A_2	A_2A_2	
A_1A_1	41	70		111
A_1A_2	65	173	119	357
A_2A_2		127	187	314
Genotypic distribution among zygotes	106	370	306	782
Observed male gametes	233		376	609

From Christiansen et al. (1973).

If all adult individuals in the population have the same chance to breed, irrespective of their genotype at the considered autosomal locus (i.e., sexual selection does not affect this gene) then the genotypic proportions in the breeding population will be the same as those in the population of adult individuals (Table 2-2). If, in addition, zygotes of the various genotypes on average have the same chance to survive to adulthood, then the genotypic proportions among zygotes and among adults will be the same. Thus, if there is no selection on this gene, the breeding individuals will occur in the same genotypic proportions as when they were formed as zygotes. In particular, the breeding

Table 2-2. *Adults and Breeders.* Data on an esterase polymorphism in *Zoarces viviparus* comparing the adult population and the breeding population of the two sexes. See Exercise 2.1.B.

	A_1A_1	A_1A_2	A_2A_2	Total
Mothers	111	357	314	782
Nonbreeding females	8	32	29	69
Adult females (total)	119	389	343	851
Adult males	54	200	177	431

	A_1	A_2	Total
Adult males	308	554	862
Male gametes	233	376	609

From Christiansen et al. (1973).

population and the zygote population will have the same allele frequencies. Thus we have established the fundamental principle of population genetics: the *law of the constancy of allele frequency*. For an autosomal gene, this law states: *If no selection affects a given gene, and if Mendelian segregation occurs, then the allele frequencies at this locus in the population will be constant from generation to generation. After one generation the allele frequencies in males and females will henceforth be the same.*

This law is nothing more than a formulation of the conservative transmission of genetic information on the level of the population. It states that, given no disturbance, the conservative transmission of genetic material holds between generations for populations as well as for individuals. Viewed in this way, the law contains the definition of the central evolutionary concept of *selection* as any force that can mediate a change in the genetic constitution of a large population.

Exercise 2.1.B

Consider the data in Table 2-2.

a. Find the allele frequencies in all the populations given in the table.
   Statistical analysis of the data shows that the differences between the mothers and the nonbreeding females, between adult females and adult males, and between adult males and male gametes may all be due to random fluctuations in finite samples (Christiansen and Frydenberg, 1973).
b. Count the number of independently observed A_1 and A_2 alleles in the total data in Tables 2-1 and 2-2. What are the gene frequencies in the total data?

2.1.2 Allele Frequencies for Sex-linked Genes

For genes on the sex chromosomes, similar results hold, although the formulation is a bit more complicated due to "criss-cross" inheritance of genes on the X-chromosome. For genes on the Y chromosome, the law of conservation of allele frequency is almost trivial, so we will only discuss the law for X-linked genes.

Consider an X-linked gene with the alleles S_1 and S_2, and suppose the genotype numbers among breeding females are G_{11}, G_{12}, and G_{22} of S_1S_1, S_1S_2, and S_2S_2, and the allele frequencies are p_1 and p_2 ($p_1 + p_2 = 1$). Then, if there is Mendelian segregation and no selection, the allele frequencies among the egg cells forming the offspring generation are p_1 and p_2. Suppose the genotypes S_1 and S_2 in males occur in the numbers H_1 and H_2 ($H_1 + H_2 = M$). Then the allele frequencies among males are

$$q_1 = \frac{H_1}{M} \text{ and } q_2 = \frac{H_2}{M},$$

with $q_1 + q_2 = 1$. Under the assumption of no selection, the allele frequencies among X-carrying sperm are q_1 and q_2. The allele frequencies among females in the zygote population are $(p_1 + q_1)/2$ and $(p_2 + q_2)/2$, since each female zygote carries an X chromosome supplied by the mother and one supplied by the father. The allele frequencies among male zygotes are p_1 and p_2 because the X chromosome carried by each male zygote is supplied by the mother. Thus if the allele frequencies were initially different in males and females, then they will remain so, and the simple conservation law that held for an autosomal gene can only hold if the allele frequencies in the two sexes are the same. In the autosomal case, however, we defined an average allele frequency in the population by giving the sexes equal weight, because each individual carries two copies of the gene and each supplies equally many alleles to the offspring. For an X-linked gene, the females carry twice as many alleles as the males and they supply copies to both male and female offspring, whereas the males supply copies only to the female offspring. Again, if males and females have equal weight, then the allele frequencies among X chromosomes in the breeding population are

$$\frac{2p_1 + q_1}{3} \text{ and } \frac{2p_2 + q_2}{3}.$$

In the same way, among the offspring zygotes the allele frequencies are

$$\frac{2(p_1 + q_1)/2 + p_1}{3} \text{ and } \frac{2(p_2 + q_2)/2 + p_2}{3}.$$

These allele frequencies are the same in the zygote population as in the breeding population. Thus the law of allele frequency conservation holds for X-linked loci, if the allele frequencies among the X chromosomes in the population are understood to have this meaning.

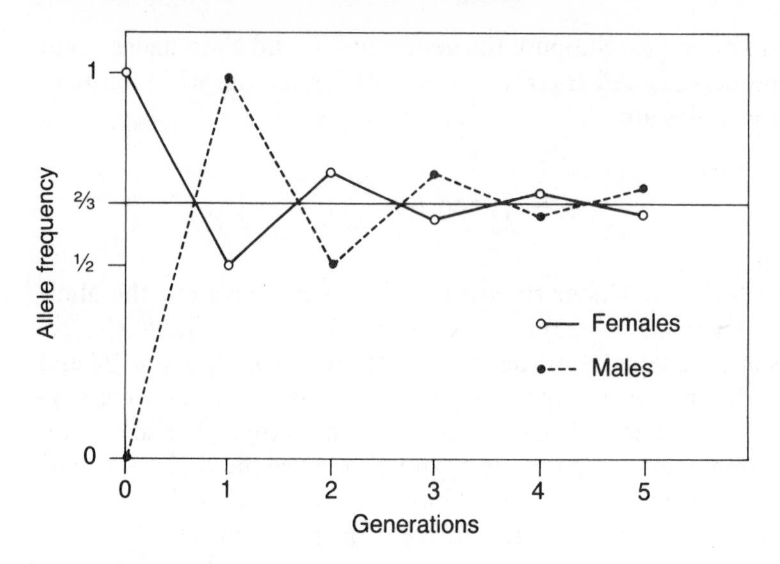

Figure 2-1. Allele frequency changes in females and males at a sex-linked locus in a population that began with the females and males homozygous for different alleles.

The initial difference between the frequency of allele S_1 in males and females (i.e., $p_1 - q_1$) becomes

$$\frac{p_1 + q_1}{2} - p_1 = \frac{-(p_1 - q_1)}{2},$$

in the next generation. Thus, the difference in allele frequencies between the sexes is halved in every generation. As time proceeds, the difference decreases, and in the limit the allele frequencies among males and females become equal, but the convergence to equality is delayed in contrast to the establishment of equality after one generation in the autosomal case. In addition, if the allele frequency of S_1 is initially higher in females than in males, then this order reverses in the next generation (Figure 2-1).

2.1.3 Finite Population Size and Random Genetic Drift

The conservation laws predict the expected value of allele frequencies in an offspring population from our knowledge of the allele frequen-

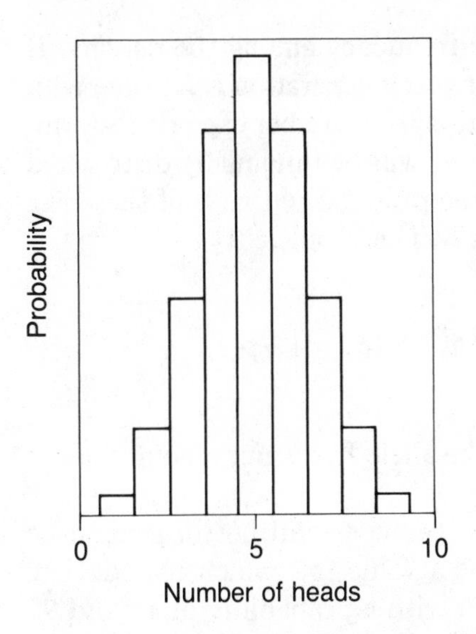

Figure 2-2. Binomial distribution with frequency parameter 1/2 and number parameter 10. This is the expected frequency distribution in a game where ten coins are tossed simultaneously and the number of heads (or tails) is counted. In a Mendelian backcross, Aa × aa, *this is the expected distribution of the number of* Aa *(or* aa*) offspring in litters of 10. Similarly, it is the expected number* A *alleles among five progeny from the* F_1*-cross* Aa × Aa.

cies in the population of parents. The laws have a stochastic interpretation similar to the laws of segregation and heredity in that they provide the probability that a randomly chosen offspring should have a given genotype. For example, a randomly chosen offspring from the cross $A_1A_1 \times A_1A_2$ has the genotype A_1A_1 with probability 1/2, but nobody would be surprised if two offspring from this cross both had the genotype A_1A_1. If we examined 10 offspring, then the probability that a given number of offspring should be A_1A_1 would be provided by the binomial distribution with parameters 10 and 1/2 (Figure 2-2). Similarly, if for some reason a parent population polymorphic for A_1 and A_2 at an autosomal locus produced only one offspring, then the gene frequency of A_1 in the offspring "population" could only be one of 0, 1/2, or 1, no matter the allele frequency among the parents. If the number of individuals in the offspring generation is N, then with Mendelian segregation and no selection, the number of egg cells forming these offspring that carry allele A_1 will be binomially distributed with parametes N and p_1; that is, the probability that k of these egg cells will carry allele A_1 and $N - k$ will carry allele A_2 is

$$b(k;p_1,N) = \binom{N}{k} p_1^k p_2^{N-k}, (p_2 = 1 - p_1)$$

This, then, is the probability that the allele frequency among the egg cells is k/N.

Thus the frequency of A_1 among the eggs will not in general be exactly equal to p_1, but will assume a value, p_1', which may deviate stochastically from p_1. For example, with a probability of around 95 percent we will find that p_1' has a value in the interval

$$p_1 - 2(p_1p_2/N)^{1/2} < p_1' < p_1 + 2(p_1p_2/N)^{1/2}.$$

In large populations this is a small range, and there is no need to worry about it when the allele frequency among offspring is predicted from the allele frequency among parents. For the law of the conservation of the allele frequency, however, these fluctuations may become important over long periods of time. The relevance of the fluctuations originates from the law itself. Let the allele frequency in a given generation be p_1; then, from the conservation law with no selection, the allele frequency in any future generation should be p_1. When the offspring population is formed, however, a small error occurs and the allele frequency is p_1', which is now the predicted allele frequency in all future generations. Thus, the small error is preserved, as there is no way in which the population can "remember" that its parents had the allele frequency p_1. Even in very large populations the allele frequency can change in time due to accumulated small random fluctuations. This change is purely accidental because in each generation the allele frequency is equally likely to increase or to decrease a bit. This chance phenomenon as known as *random genetic drift,* and its importance in population genetics was originally stressed by the great American population geneticist Sewall Wright.

After a long time the population will lose either of the alleles A_1 or A_2, just as playing a perfectly honest game of coin tossing produces a winner and a broke loser after sufficiently long time. Thus, a population with A_1 initially at frequency p_1 (< 1) will become monomorphic A_1A_1 or A_2A_2, after which no further change in the allele frequency can occur. From our conservation law, however, we would expect that the gene frequency would stay equal to p_1. In this context this means that if there were a lot of populations, all starting with gene frequency p_1, then the average gene frequency among the populations would remain p_1. If after a very long time all the populations were monomorphic, then a fraction p_1 should be monomorphic A_1A_1 and a fraction p_2 should be monomorphic A_2A_2. In other words, given enough time, the population becomes monomorphic A_1A_1 with probability p_1.

How long is "enough" time? It turns out that after on the order of N generations, where N is the population size, random genetic drift plays a major role in displacing the gene frequency (Figure 2-3). In a human population of 10,000, this would be on the order of 300,000 years; only in very small populations will random genetic drift become important within a time span of, say, the history of civilization. The population size referred to here, however, is that of the breeding population, which may deviate from the actual head count. In addition, variation among individuals in their number of offspring adds to the probable error in transmission of allele frequency between generations. The population size N in the preceding evaluation may therefore deviate from the count of the breeding individuals. This number, the *effective population size,* can be evaluated from demographic data on the population (see Kimura and Ohta, 1971, or Ewens, 1979).

In conclusion, for large populations we can consider the laws of allele frequency conservation to be exact over quite long time spans. For very small populations the laws must be viewed in the sense of expectation, allowing for random fluctuations (Figure 2-4). Over sufficiently long time spans, as determined by the effective population size, all finite populations must be considered as "small" populations.

So far we have only considered two alleles at a locus. These results easily generalize to a situation where k alleles, $A_1, A_2, \ldots, A_k$, are present in the population. To see this, just consider allele A_1 and the collective allele A' defined as the alleles of type $A_2, A_3, \ldots,$ or A_k. This takes us back to the two-allele case, so all results hold for allele A_1. Similar operations with the other alleles validate the generalization to multiple alleles of the segregation law, the law of heredity, the law of the conservation of the allele frequency, and the results pertaining to random genetic drift. Therefore, if the allele frequencies of $A_1, A_2, \ldots, A_k$ are $p_1, p_2, \ldots, p_k$ ($p_1 + p_2 + \ldots + p_k = 1$), then in a large

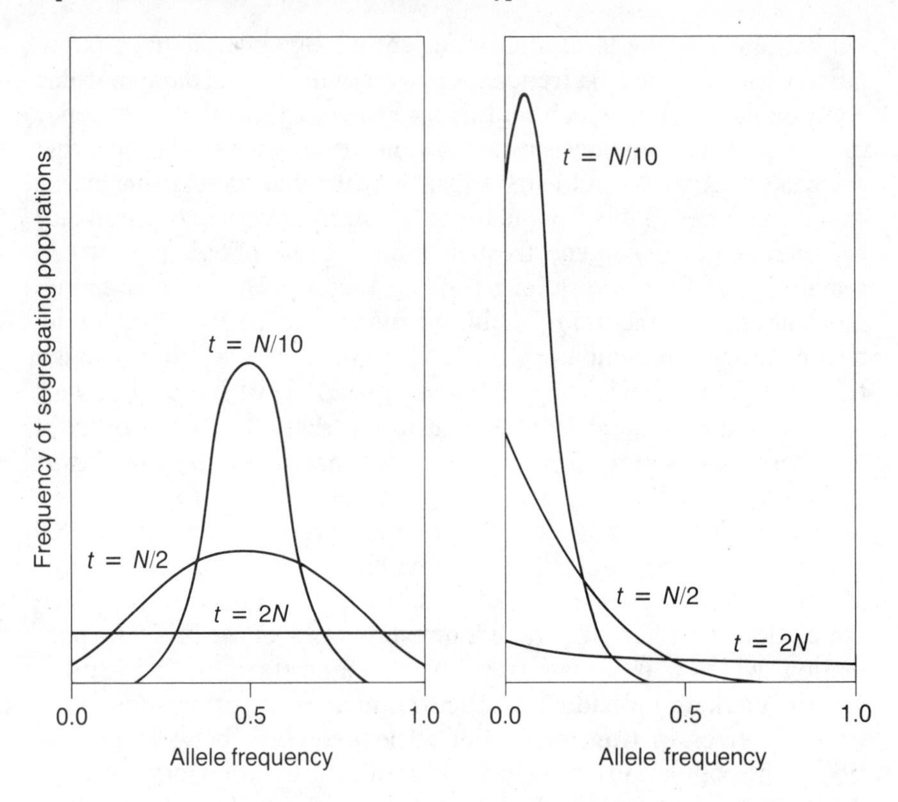

Figure 2-3. Left: *Divergence of gene frequencies due to random genetic drift. Imagine a large number of populations all of size* N *(i.e., each having* 2N *genes at a given locus). Each population starts with an allele frequency of 1/2, and we follow the allele frequency in all populations through time. After* N/*10 generations, the allele frequencies will have the distribution shown at the left. The allele frequencies still cluster around 0.5, but a small fraction of the populations may have allele frequencies as low as 0.2 or as high as 0.8. After* N/*2 generations, populations with all gene frequencies are possible, although some clustering around 0.5 is still apparent. Now, however, there are some very low frequencies and some very high frequencies; that is, fixation at allele frequencies at 0 and 1 occurs. The figure does not give the frequency of the fixed populations, only the distribution among populations that are still segregating. After* 2N *generations, the distribution of allele frequencies among segregating populations will be flat. This shows that after such a length of time a segregating population is equally likely to have any possible allele frequency.* Right: *The same process is illustrated in populations in which the initial allele frequency was 0.1. Here fixation of populations occurs earlier, but otherwise the picture of the process of random genetic drift is the same. (After Kimura, 1955.)*

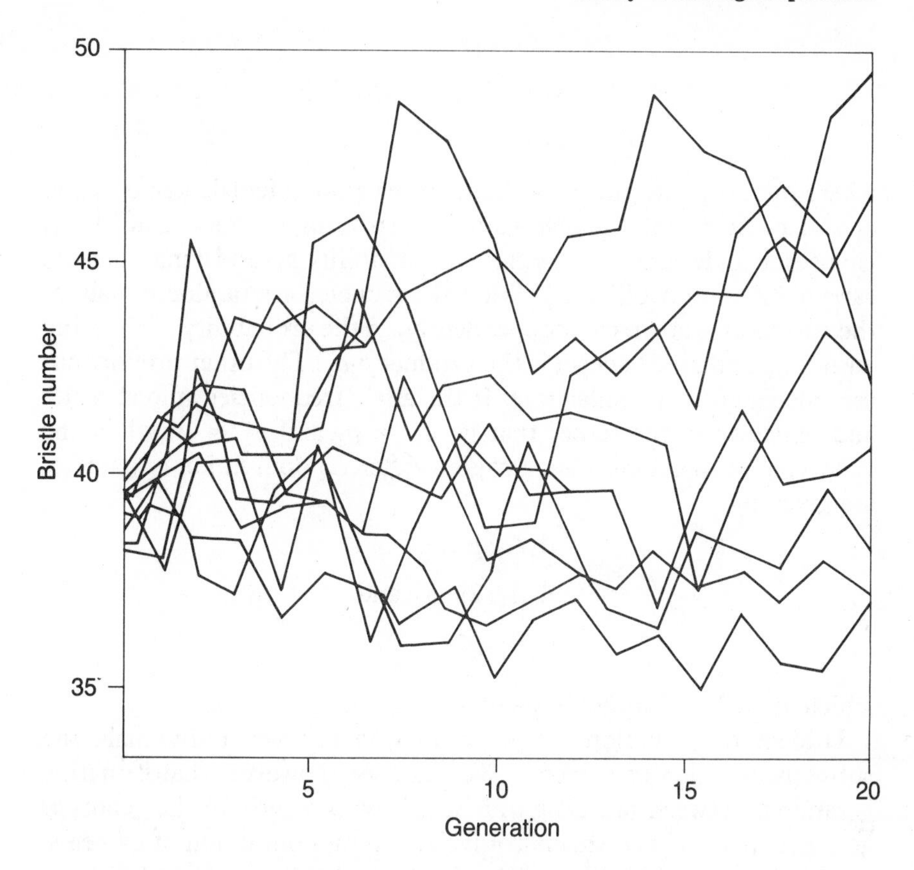

Figure 2-4. Divergence for a bristle number character among 10 populations of Drosophila melanogaster, *each consisting of one female and one male every generation. (After Rasmuson, 1952.)*

population these allele frequencies will be conserved in all future generations if no selection acts on the considered locus.

2.2 Hardy-Weinberg Proportions

Genes are transmitted conservatively from parent to offspring or from parent population to offspring population under the assumption of no selection. Genotypes, however, are not transmitted. Genotypic proportions are determined from the way in which the male and female gametes unite to form the zygotes. If the gametes unite at random, the genotypic proportions among zygotes at an autosomal locus with the alleles A_1 and A_2 are

$$\text{for } A_1A_1\colon p_1q_1,$$
$$\text{for } A_1A_2\colon p_1q_2 + p_2q_1, \text{ and}$$
$$\text{for } A_2A_2\colon p_2q_2,$$

where p_1 and p_2 are the allele frequencies among female gametes and q_1 and q_2 those among male gametes. To form a homozygote A_1A_1, choose a female gamete carrying A_1 (probability p_1) and a male gamete carrying A_1 (probability q_1). Since the gametes are randomly paired, the two choices are made independently, so the probability of choosing an A_1 egg and an A_1 sperm is the product p_1q_1. The other proportions are calculated in a similar way. If the allele frequencies among males and females are the same, that is, $q_1 = p_1$ and $q_2 = p_2$, then the genotypic proportions among zygotes after random union of gametes are given by

$$A_1A_1\colon p_1^2,$$
$$A_1A_2\colon 2p_1p_2, \text{ and}$$
$$A_2A_2\colon p_2^2,$$

which are called *Hardy-Weinberg proportions*.

Human reproduction occurs by mating between individuals and subsequent union of gametes. We shall see, however, that if mating is random between breeding individuals with respect to the genotype at a given locus, then this is equivalent to random union of gametes, in the absence of both gametic and fecundity selection. Choose an offspring at random; this selects a random pair of parents with respect to their genotype at a given locus since no fecundity selection occurs. Both male and female gametes from these parents were chosen at random since there is no gametic selection. Thus, we have chosen a random male gamete from among the gametes produced by the breeding male population and united it with a random female gamete from among the gametes produced by the population of breeding females (since the pair of parents was formed by random mating). The randomly chosen offspring was therefore formed by the union of randomly chosen male and female gametes, by random union of gametes.

The genotypic proportions that were obtained from random union of gametes can be obtained directly from the random mating frequencies. Let $g_{11} = G_{11}/N$, $g_{12} = G_{12}/N$, and $g_{22} = G_{22}/N$ be the genotypic proportions among breeding females, and let $h_{11} = H_{11}/M$, $h_{12} = H_{12}/M$ and $h_{22} = H_{22}/M$ be those among breeding males. The genotypic frequencies among the offspring are calculated from Table 2-3.

Table 2-3. *Results of Random Mating.*

Mating	Frequency	Frequency of offspring		
♀ × ♂		A_1A_1	A_1A_2	A_2A_2
$A_1A_1 \times A_1A_1$	$g_{11}h_{11}$	$g_{11}h_{11}$		
$A_1A_1 \times A_1A_2$	$g_{11}h_{12}$	$\frac{g_{11}h_{12}}{2}$	$\frac{g_{11}h_{12}}{2}$	
$A_1A_1 \times A_2A_2$	$g_{11}h_{22}$		$g_{11}h_{22}$	
$A_1A_2 \times A_1A_1$	$g_{12}h_{11}$	$\frac{g_{12}h_{11}}{2}$	$\frac{g_{12}h_{11}}{2}$	
$A_1A_2 \times A_1A_2$	$g_{12}h_{12}$	$\frac{g_{12}h_{12}}{4}$	$\frac{g_{12}h_{12}}{2}$	$\frac{g_{12}h_{12}}{4}$
$A_1A_2 \times A_2A_2$	$g_{12}h_{22}$		$\frac{g_{12}h_{22}}{2}$	$\frac{g_{12}h_{22}}{2}$
$A_2A_2 \times A_1A_1$	$g_{22}h_{11}$		$g_{22}h_{11}$	
$A_2A_2 \times A_1A_2$	$g_{22}h_{12}$		$\frac{g_{22}h_{12}}{2}$	$\frac{g_{22}h_{12}}{2}$
$A_2A_2 \times A_2A_2$	$g_{22}h_{22}$			$g_{22}h_{22}$
Total	1	p_1q_1	$p_1q_2 + p_2q_1$	p_2q_2

If no selection acts on this gene, and if there is random mating, then the allele frequencies are constant from generation to generation, they are the same in the two sexes, and the genotypic frequencies are in Hardy-Weinberg proportions. This state of the population is known as *Hardy-Weinberg equilibrium.*

Table 2-3 can be simplified when the population is in Hardy-Weinberg equilibrium, in which case the sex of the parent is immaterial and the genotypic proportions in the breeding population are Hardy-Weinberg proportions, as in Table 2-4.

In addition, a table of combinations of one parent, say, the mother with an offspring, can be obtained by summing over the male parental genotypes, as in Table 2-5. From Table 2-5 random mating between genotypes implies that the offspring of a given mother is formed by the union of one of her gametes with a randomly chosen sperm in the population (cf. Table 2-1).

Exercise 2.2.A

Consider the data in Table 2-1 and the answer to Exercise 2.1.A.

a. Is it reasonable to infer random mating in this population?

Table 2-4. *Random Mating with Hardy-Weinberg Equilibrium.*

Mating	Frequency	Frequency of offspring		
		A_1A_1	A_1A_2	A_2A_2
$A_1A_1 \times A_1A_1$	p_1^4	p_1^4		
$A_1A_1 \times A_1A_2$	$4p_1^3p_2$	$2p_1^3p_2$	$2p_1^3p_2$	
$A_1A_1 \times A_2A_2$	$2p_1^2p_2^2$		$2p_1^2p_2^2$	
$A_1A_2 \times A_1A_2$	$4p_1^2p_2^2$	$p_1^2p_2^2$	$2p_1^2p_2^2$	$p_1^2p_2^2$
$A_1A_2 \times A_2A_2$	$4p_1p_2^3$		$2p_1p_2^3$	$2p_1p_2^3$
$A_2A_2 \times A_2A_2$	p_2^4			p_2^4
Total	1	p_1^2	$2p_1p_2$	p_2^2

b. Compare the genotypic frequencies among mothers and among offspring to the Hardy-Weinberg proportions.

c. Compare the data in Table 2-2 to the Hardy-Weinberg proportions.

Consider now an autosomal locus with multiple alleles, say, A_1, A_2, ..., A_k with no distinction between the sexes and assume that no selection acts. Then the allele frequencies $p_1, p_2, \ldots, p_k$ are conserved from generation to generation. With random mating, the genotypic proportions are then

$$\text{for homozygote } A_iA_i\text{: } p_i^2 \text{ and}$$

$$\text{for heterozygote } A_iA_j\text{: } 2p_ip_j,$$

for all $i = 1, 2, \ldots, k$ and $j = 1, 2, \ldots, k$ with $i \neq j$. These proportions can be found by the same arguments as for two alleles, either by using the equivalence of random mating and random union of gametes, or by constructing the mating table.

For a sex-linked locus with alleles S_1 and S_2 at frequencies p_1 and p_2 in females and q_1 and q_2 in males, the random mating genotypic proportions among their offspring are

$$\text{for } A_1A_1 \text{ females: } p_1q_1,$$

$$\text{for } A_1A_2 \text{ females: } p_1q_2 + p_2q_1, \text{ and}$$

$$\text{for } A_2A_2 \text{ females: } p_2q_2.$$

Table 2-5. *Mother-offspring Combinations.*

Mother Genotype	Frequency	Frequency of offspring			Frequency of offspring within mother type		
		A_1A_1	A_1A_2	A_2A_2	A_1A_1	A_1A_2	A_2A_2
A_1A_1	p_1^2	p_1^3	$p_1^2p_2$		p_1	p_2	
A_1A_2	$2p_1p_2$	$p_1^2p_2$	p_1p_2	$p_1p_2^2$	$\frac{p_1}{2}$	$\frac{1}{2}$	$\frac{p_2}{2}$
A_2A_2	p_2^2		$p_1p_2^2$	p_2^3		p_1	p_2
Total	1	p_1^2	$2p_1p_2$	p_2^2			

Since the males are hemizygous, the genotypic proportions are

$$\text{for } A_1 \text{ males: } p_1, \text{ and}$$

$$\text{for } A_2 \text{ males: } p_2.$$

If the allele frequencies among males and females are equal (as they will be after a long time if no selection occurs), then the female genotypes will be in Hardy-Weinberg proportions and the male genotype frequencies will be equal to the allele frequencies.

Exercise 2.2.B (Theoretical)

Show that the frequency of heterozygotes in a two-allele polymorphism in the zygote population is always higher than that expected from the Hardy-Weinberg proportions if the allele frequencies among parents differ between the sexes.

What does this say about the genotypic proportions among females at an X-linked locus?

Exercise 2.2.C

Consider a human population polymorphic at an autosomal locus for alleles A_1 and A_2 with allele frequencies p_1 and p_2, and assume that mating is random with respect to this variation and that no selection occurs.

a. Among families with two children, find the frequencies of the nine combinations of genotypes of the sibs when the genotypes of the parents are unknown, that is, fill out the following table:

Genotype of first child	Genotype of second child			Total
	A_1A_1	A_1A_2	A_2A_2	
A_1A_1				
A_1A_2				
A_2A_2				
Total				

b. What are the genotypic proportions among the firstborn children of these families? Use this to check the totals in the rightmost column.
This exercise requires a fair amount of work.

2.2.1 Phenotypic Proportions

For loci where the different genotypes can be observed as phenotypes (e.g., via simple electrophoretically defined protein differences), the allele frequencies can be inferred directly from a sample by counting the observed alleles. However, if the number of observable phenotypes is fewer than the number of genotypes because of dominance, then the allele frequencies can usually only be inferred if something is known about the genotypic proportions, that is, about the matings.

If we suppose that random mating and no selection occur, then the genotypic proportions at an autosomal locus will be the Hardy-Weinberg proportions. Suppose there are two alleles, A and a, with A dominant and a recessive, then two observable phenotypes exist, namely, the dominant phenotype A− and the recessive phenotype aa. If the allele frequency of A is p and that of a is q, then the phenotypic proportions are

$$\text{for A}-: d = p^2 + 2pq = 1 - q^2, \text{ and}$$

$$\text{for aa: } r = q^2.$$

The allele frequency q of a might then be estimated by the square root of the frequency of the recessive phenotype. However, there is no way in which we can check the assumption of random mating from the data on the phenotypic frequencies in the population, as any pair of observed phenotype frequencies, d and r, correspond to the Hardy-Weinberg proportions for some pair of gene frequencies. One way to check the assumption of random mating is by observing the matings in the population. Another way is to check that the adult genotypes are in Hardy-Weinberg proportions, and this can be done by observing the frequencies of the alleles among gametes from the dominant phenotype in parent-offspring combinations. From the assumption of Hardy-Weinberg proportions, the heterozygote *Aa* has the frequency

$$\frac{2pq}{p^2 + 2pq}$$

among the fraction of the population that shows the dominant phenotype. The homozygote *AA* segregates no gametes carrying *a* and the heterozygote *Aa* segregates half of its gametes as *a*. Thus, among individuals of the dominant phenotype a fraction

$$S = \frac{pq}{p^2 + 2pq} = \frac{q}{1 + q}$$

of the gametes will carry allele *a*. This segregation ratio is called *Snyder's ratio*. The offspring segregation proportions for the three types of phenotypic matings are then as in Table 2-6. The actual segregation among offspring phenotypes in any mating is Mendelian, but the average segregation among offspring phenotypes in matings between given phenotypes is as given by Table 2-6. Another more indirect indication of the presence of Hardy-Weinberg proportions can be obtained from mother-offspring combinations, which are recorded in Table 2-7 for the case of dominance. Thus if the genotypes among females are in Hardy-Weinberg proportions, then the frequency of aa children from A− mothers is a fraction S, Snyder's ratio, of the frequency of aa children from aa mothers.

Table 2-6. *Mating Table in the Case of Dominance.*

Mating	Segregation among offspring within mating type		Mating frequency
	A−	aa	
A− × A−	$1 - S^2$	S^2	d^2
A− × aa	$1 - S$	S	$2dr$
aa × aa	0	1	r^2

Table 2-7. *Mother-offspring Table in the Case of Dominance.*

Mother phenotype	Segregation among offspring within mother phenotype		Mother frequency
	A−	aa	
A−	$1 - qS$	qS	$p(1 + q)$
aa	p	q	q^2

Exercise 2.2.D

Human blood is polymorphic with respect to the reaction with P-antiserum (positive reaction: +; negative reaction: −). The frequencies of the two phenotypes are equal in males and females. The table shows how the P-blood types are inherited:

Parents	Number	Children +	Children −	Total
Both +	249	677	79	756
One + and one −	174	349	212	561
Both −	34	0	94	94
Total	457	1026	385	1411

a. How are the P-blood types inherited?

b. Is it reasonable to infer random mating among the parents?

Assuming random mating and no selection:

c. Find the allele frequencies among parents.

d. What are the expected frequencies of the two phenotypes among children from the three types of parental pairs? Are these frequencies in reasonable agreement with the observations?

Exercise 2.2.E

In an investigation of the Danish population, 5500 persons were typed for reaction to the Rhesus D-antiserum: 4586 showed a positive reaction and 914 a negative reaction (Gürtler and Henningsen, 1954). Rhesus-positive is dominant to Rhesus-negative and the inheritance is autosomal. In the following, assume that the genotype frequencies are in the Hardy-Weinberg proportions.

a. What are the allele frequencies in the Danish population?

b. Find the genotype frequencies within the Rhesus-positive phenotype.

The Rhesus-phenotype defined by the reaction with D-antiserum is involved in the mother-child incompatibility reaction that may cause the Rhesus disease (*Erythroblastosis foetalis*) in newborn babies. The risk of the Rhesus disease is present whenever a Rhesus-negative mother carries a Rhesus-positive fetus, and the risk increases with the number of Rhesus-positive children born to the Rhesus-negative mother. To form an impression of these risks, calculate the probabilities of the following events in the Danish population.

c. A Rhesus-negative woman marries a Rhesus-positive man.

d. A Rhesus-positive man is homozygote; is heterozygote.

e. A Rhesus-negative woman whose husband is Rhesus-positive bears a Rhesus-positive fetus in her first pregnancy.

Exercise 2.2.F

Consider a human population polymorphic for two phenotypes known to be determined by the dominant allele A and the recessive a at an autosomal locus. Among families with two children find the frequencies of the four phenotypic combinations of the sibs assuming that the phenotypes of the parents are unknown and that the population is in Hardy-Weinberg equilibrium.

Phenotype of first child	Phenotype of second child	
	A–	aa
A–		
aa		

Exercise 2.2.G

The allele frequencies at the ABO locus in Europe are 0.27, 0.08, and 0.65 for the alleles corresponding to blood types A (blood types A_1 and A_2 combined), B, and O, respectively (Table 1-8). The alleles for blood types A and B are codominant and the allele for blood type O is recessive.

a. Calculate the genotype frequencies at this locus assuming Hardy-Weinberg equilibrium.

b. What are the phenotypic frequencies (blood type frequencies) in the population, again assuming Hardy-Weinberg equilibrium?

c. What is the probability that a blood type O mother bears a fetus with blood type A, B, or AB (cf. section 1.2.3)?

d. What is the probability that the first child is blood type O from a mating of parents both with blood type A?

2.3 Two-Locus Gametic Proportions

The description of genetic variation and the transmission of genetic information on the population level begins with consideration of allelic variation at a single gene locus. The goal, however, is to describe the variation in the whole genome with thousands of interconnected gene loci. Some of the new phenomena involved in this extension are apparent even when the variation is considered only at two gene loci simultaneously. Consider two autosomal loci A and B, each with two alleles, A_1, A_2 and B_1, B_2, respectively, and assume that no selection acts on either gene so that the allele frequencies at each locus are preserved.

The four types of gametes, A_1B_1, A_1B_2, A_2B_1, and A_2B_2, can be united to form zygotes in ten different ways so that in general ten different two-locus genotypes have to be considered. If we inspect the genotypes at each locus separately, however, only nine different types of individuals actually occur, because individuals that are heterozygotes at both loci can be formed by uniting either an A_1B_1 gamete with an A_2B_2 gamete or an A_1B_2 gamete with an A_2B_1 gamete. In the following we will distinguish the frequencies of these two double heterozygote genotypes even in the case where the loci are unlinked.

Exercise 2.3.A

The genotypes of all the offspring of a *Zoarces viviparus* female of genotype $EstIII^1/EstIII^2$ at an esterase locus and $PgmI^1/PgmI^2$ at a phosphoglucomutase locus were determined at both of these loci with the following result (Simonsen and Frydenberg, 1972):

	$PgmI^1/PgmI^2$	$PgmI^2/PgmI^2$
$EstIII^1/EstIII^2$	51	48
$EstIII^2/EstIII^2$	56	46

a. What is the most likely genotype of the father of these offspring?

b. What is the recombination frequency between these loci?

The new phenomenon of genetic transmission that enters here is *recombination*. Suppose the recombination frequency between the two loci in both sexes is c, $0 \leq c \leq 1/2$, and consider an individual of genotype $A'B'/A''B''$ formed by uniting the gametes $A'B'$ and $A''B''$, where A' is either of the alleles A_1 or A_2, B' is either of B_1 or B_2, and so on. When this individual produces gametes, a fraction $1 - c$, the *parental gametes,* are either $A'B'$ or $A''B''$, and the remaining fraction c, the *recombinant gametes,* are either $A'B''$ or $A''B'$. We have assumed Mendelian segregation, so the four gametes are produced in the proportions

$$\frac{1-c}{2}\mathrm{A'B'} : \frac{c}{2}\mathrm{A'B''} : \frac{c}{2}\mathrm{A''B'} : \frac{1-c}{2}\mathrm{A''B''}.$$

Recombination entails an interaction of the two gametes that formed the individual, so the gametic output from a breeding population is dependent on its genotypic composition. Therefore, the process of recombination can only be studied by assuming a mating rule for the population. To simplify the arguments, we will only study the process of recombination in a population where the zygotes are formed by *random union of gametes.*

Let x_{11}, x_{12}, x_{21}, and x_{22} ($x_{11} + x_{12} + x_{21} + x_{22} = 1$) be the frequencies of the four gametes A_1B_1, A_1B_2, A_2B_1, and A_2B_2 among the gametes produced by the breeding individuals in a given generation. The allele frequencies among these gametes are

$$p_{A_1} = x_{11} + x_{12},\ p_{A_2} = x_{21} + x_{22},$$

$$p_{B_1} = x_{11} + x_{21},\ p_{B_2} = x_{12} + x_{22}.$$

The offspring population is formed by uniting these gametes at random into zygotes, so the genotypic proportions among the zygotes are given by the Hardy-Weinberg proportions corresponding to four gametic types in the frequencies x_{11}, x_{12}, x_{21}, and x_{22}. The goal is to find the frequencies x'_{11}, x'_{12}, x'_{21}, and x'_{22} among the gametes produced by this offspring population. We have assumed no selection so the allele frequencies are conserved. That is,

$$x'_{11} + x'_{12} = p_{A_1},\ x'_{21} + x'_{22} = p_{A_2},$$

$$x'_{11} + x'_{21} = p_{B_1},\ x'_{12} + x'_{22} = p_{B_2},$$

Thus we have only to determine x'_{11} to provide all four gametic frequencies.

To ease the calculation of x'_{11}, divide the gametes produced in the offspring generation into the fraction $1 - c$ of parental gametes and the fraction c of recombinant gametes. The parental gametes are unchanged copies of the gametes produced by the parental generation, so the frequency of A_1B_1 gametes among these is x_{11}. The recombinant gametes contain a copy of the A locus allele from one of the gametes that formed the individual and a copy of the B allele from the other gamete. The frequencies of gametes that carry alleles A_1 and B_1 are p_{A_I} and p_{B_I}, respectively. With random union of gamete proportions in the breeding population, the frequency of A_1B_1 gametes among recombinant gametes is $p_{A_1}p_{B_1}$. Thus we have shown that

$$x'_{11} = (1 - c)x_{11} + cp_{A_1}p_{B_1}.$$

From this expression it follows that the gamete frequency will change from generation to generation when recombination occurs, that is, when c $\neq 0$. However, if the gamete frequencies happen to be given by the product of the allele frequencies: $x_{11} = p_{A_1}p_{B_1}$, then no change in the gamete frequencies occurs. The population is said to be at equilibrium due to recombination, or in *linkage equilibrium*. When $x_{11} = p_{A_1}p_{B_1}$, then it is easy to see that $x_{12} = p_{A_1}p_{B_2}$, $x_{21} = p_{A_2}p_{B_1}$, and $x_{22} = p_{A_2}p_{B_2}$, and all the gametic frequencies have a characteristic product form, the *Robbins proportions*. The gametic proportions converge to these proportions as seen by rewriting the previous equation as

$$x'_{11} - p_{A_1}p_{B_1} = (1 - c)(x_{11} - p_{A_1}p_{B_1}).$$

Thus the deviation of the gametic proportions from Robbins proportions is a factor $(1 - c)$ smaller than the deviation in the parental population. The deviation from Robbins proportions is called the *linkage disequilibrium*

$$D = x_{11} - p_{A_1}p_{B_1}.$$

Hence $D' = (1 - c)D$ in the offspring generation, and after n generations we have $D^{(n)} = (1 - c)^n D$. Thus, the linkage disequilibrium converges to zero with time as $(1 - c)^n$, and for any recombination frequency in the interval $0 < c \leqslant 1/2$ the population will eventually have gametic frequencies in Robbins proportions, although the time needed to reach that equilibrium may be long if c is very small.

The delay in the reshuffling of the alleles in the gametes occurs because changes in the gamete frequencies only occur by recombination in the double heterozygotes. The linkage disequilibrium can be rewritten as

$$D = x_{11}x_{22} - x_{12}x_{21},$$

which is half the difference between the genotypic frequencies of the two double heterozygotes when the genotypic frequencies are in Hardy-Weinberg proportions. If random mating cannot be assumed, this difference in genotypic proportions again plays a decisive role in the evolution of the population. This general theory for two loci is complicated, however, and will not be considered here.

The Robbins proportions correspond to independent distribution of the alleles at the two loci into the gametes. Thus, when linkage equilibrium is reached, the two loci can be considered as independently transmitted on the population level no matter what their actual linkage relationship. On the other hand, even if the loci are unlinked ($c = 1/2$) and the gametic frequencies deviate for some reason from Robbins proportions, they will take time to achieve these proportions, with the linkage disequilibrium being halved every generation. In conclusion, the association between two loci on the population level, as measured by the linkage disequilibrium, bears no immediate relationship to formal genetic linkage; tightly linked loci are expected to be independently transmitted on the population level after a sufficiently long time, and unlinked loci may show episodes of linkage *on the population level* by virtue of their being in linkage disequilibrium.

Exercise 2.3.B

Consider two autosomal loci with alleles A,a and B,b on the same chromosome. Assume random mating and no selection, and suppose the population frequency of *aa* is 0.81, of *bb* 0.81, and of *aa bb* $0.6561 = (0.81)^2$. We are interested in matings of double heterozygotes with individuals of genotype *aa bb,* because the genotypic proportions among offspring of this mating provide information on the recombination frequency between the loci. The mating is a linkage test cross.

a. What is the frequency of this kind of mating in the population?
b. Try the same question when the frequency of *aa* is 0.64, of *bb* 0.49 and of *aa bb* 0.25.

Exercise 2.3.C

Consider two loci such that the gametic frequencies are A_1B_1: 0.5, A_1B_2: 0.1, A_2B_1: 0.2, and A_2B_2: 0.2.

a. Calculate the gene frequencies at the two loci, the corresponding Robbins proportions, and the linkage disequilibrium.

The frequency of recombination between the loci is 0.05.

b. How long will it take for the linkage disequilibrium to reach half the initial value?

Exercise 2.3.D (Theoretical)

a. In the two-locus, two-allele model of section 2.3, what are the maximum and minimum values that the linkage disequilibrium can take with given allele frequencies at the two loci?

A measure of association between the alleles at two loci called the *gametic correlation* is often used. It is given by

$$\delta = \frac{D}{(p_{A_1}p_{A_2}p_{B_1}p_{B_2})^{1/2}}.$$

b. Show that $-1 \leq \delta \leq 1$.

A commonly occurring departure from the assumptions made here is sex difference in the recombination frequency. It is not difficult, however, to modify this argument to produce the result that the gamete frequencies in the two sexes become equal after one generation of random mating, and the linkage disequilibrium obeys the relation

$$D' = \left(1 - \frac{c_{♀} + c_{♂}}{2}\right)D.$$

Hence the previous results hold if c is substituted by the mean of the recombination frequencies in the two sexes.

Exercise 2.3.E (Theoretical, difficult)

Suppose that the recombination frequencies in males and females are different, $c_{♀} \neq c_{♂}$. Suppose also that reproduction is by random union of gametes.

a. Will the zygotic frequencies be in Hardy-Weinberg proportions?

b. When will the zygotic frequencies be in Hardy-Weinberg proportions?

c. (Not easy!) How would you quantify linkage disequilibrium in this population?

These results for two locus gametic proportions hold in large populations in a sense similar to that for the validity of the conservation of allele frequencies in large populations. Even in large populations, however, random genetic drift may disturb the convergence to linkage equilibrium for very closely linked loci. If we consider a breeding population formed from gametes with frequencies in Robbins proportions, then this population is expected to produce gametes in Robbins proportions with the same allele frequencies. However, the gametic frequencies among the finite number of gametes that form the offspring population will deviate a bit from these expected proportions. This deviation is expected to produce a small random change in allele frequencies among offspring, but it is also expected to produce a small random deviation from the Robbins proportions with these new allele frequencies. Thus a small deviation from zero of the linkage disequilibrium D will usually occur. When this offspring population produces gametes, a fraction c of the linkage disequilibrium will be eliminated, and the linkage disequilibrium among these gametes will be $(1 - c)D$. When c is small, the random deviation from Robbins proportions is almost conserved, with the consequence that random genetic drift may cause the accumulation of linkage disequilibrium through time. As was the case for random genetic drift of allele frequencies, this accumulation of linkage disequilibrium is purely accidental, in that a large number of populations started with the same gamete frequencies will on the average have zero linkage disequilibrium.

The natural time scale in random genetic drift is N generations, where N is the effective population number (section 2.1.3). If the convergence to linkage equilibrium occurs on the same time scale or slower, that is, if $c < 1/N$, then the stochastic buildup of linkage disequilibrium becomes significant. Thus, if the number of recombinant gametes, $2cN$, produced per generation is less than 1, then deviation from linkage equilibrium is expected to occur. Indeed, if recombination is very rare, $2cN << 1$, the deviation from linkage equilibrium is expected to be maximal, that is, the population will typically lack at least one of the four gametes. On the other hand, if recombination is frequent, $2cN >> 1$, then stochastic deviations in the linkage disequilibrium can be safely ignored.

Among the examples in Chapter 1, some of the associations of alleles at the various HLA loci (although not all of them) may be examples of linkage disequilibrium due to random genetic drift (section 1.2.4), since the recombination frequency in the HLA region is indeed very low. As the recombination process is very slow in real time, however, other historical phenomena may play a role in the HLA associations (see section 2.4.3).

Exercise 2.3.F

The material in Exercises 1.2.A and 2.2.C provides the following joint phenotypic distribution for Rhesus C and Rhesus D blood types:

	CC	Cc	cc	Total
D−	1012	2710	864	4586
dd	1	54	859	914
Total	1013	2764	1723	5500

From Gürtler and Henningsen (1954).

Are these two Rhesus loci in linkage equilibrium in the Danish population?

2.4 Deviations from Random Mating

In human populations, as well as in populations of other animals and plants, the assumption of random mating made in section 2.2 may be violated. The mating may be random with respect to individual genotypes or phenotypes, but there may be a tendency to choose mates among family members, typically with some limits to the relatedness. This inbreeding, when viewed at the population level, introduces a correlation between the genotypes of the mating pairs, which is effectively preferential mating between like genotypes. A similar deviation from random mating occurs when individuals choose mates according to phenotype, as in assortative mating (see section 1.1.3.1).

Deviations from random mating may also occur if mate choice is performed within subgroups of the population (see again section 1.1.3.1). An obvious example would be a population in which mate choice is influenced by the geographical distance between individuals, that is, a geographically structured population. The resulting genotypic similarity between mates is just one of a number of important consequences of geographic structure in a population.

2.4.1 Inbreeding

One cause of a deviation from random mating, common in some human populations, is preferential mating of related individuals (Table 2-8). Offspring of such consanguineous matings can be homozygous for alleles

Table 2-8. *Inbreeding in Human Populations.* Frequencies of consanguineous matings in human populations with F values calculated as mean coefficients of coancestry of consanguineous matings with coefficient of coancestry larger than 1/64. F is defined on page 90.

Country	Period	Uncle-niece/ aunt-nephew	First cousins (sibs)	First cousins (half sibs)	F
Belgium	1918–59	0.0002	0.0049	0.0016	0.0005
Italy	1911–15	0.0005	0.0162	0.0048	0.0015
	1956–60	0.0001	0.0077	0.0023	0.0007
U.S.A.					
Roman Catholics	1959–60		0.0008	0.0002	0.00009
Mormons	1920–40		0.0061		0.00038
India					
Andra-Pradesh	1957–58	0.0923	0.3330		0.032

After Cavalli-Sforza and Bodmer (1971).

of a given gene because the DNA of that gene in the germ cells of both parents are copies of a single DNA molecule in a common ancestor. Two such genes are said to be *identical by descent,* and an individual who receives two identical copies of a gene from its parents is said to be an identical homozygote for that gene. All individuals whose parents are related to the same extent have the same nonzero probability of being an identical homozygote for *any* autosomal gene, and these individuals are called *inbred.*

The degree of inbreeding of an individual I is measured by the probability that it is an identical homozygote for a given autosomal gene, which is called the *inbreeding coefficient* f_I. The relationship between two individuals K and L is described by the probability that two randomly chosen gametes, one from each, carry genes identical at a given autosomal locus. This is called the *coefficient of coancestry* f_{KL}. Accordingly, the coefficient of inbreeding of an individual equals the coefficient of coancestry of its parents, that is, in the pedigree

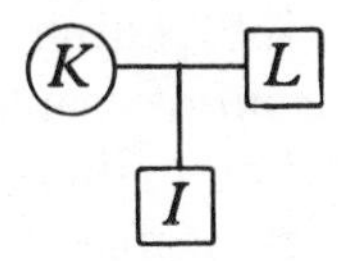

we have $f_I = f_{KL}$. Unrelated individuals have a coefficient of coancestry of zero, and the coefficient of coancestry for related individuals where the pedigree is known can be calculated by using the rules of Mendelian segregation (as reviewed in Malecot, 1969). Computations for some simple pedigrees are given in Table 2-9.

Exercise 2.4.A

a. Show that the coefficient of coancestry between mother and offspring is 1/4 when neither the mother nor the offspring are inbred.
b. What is the coefficient of coancestry of a noninbred individual with itself?
c. Show that the coefficient of coancestry between noninbred sibs is 1/4.
d. Show that the coefficient of coancestry between an inbred mother and a noninbred offspring is 1/2 at a locus where the mother is homozygous for identical alleles.
e. Show that the coefficient of coancestry between an inbred mother and a noninbred offspring is $(1 + F)/4$, where F is the inbreeding coefficient of the mother.

In a population with the autosomal alleles $A_1, A_2, \ldots, A_k$ in the frequencies $p_1, p_2, \ldots, p_k$, identical homozygotes can be of k different types, and among the identical homozygotes, A_iA_i has the frequency p_i. If mating is random with respect to the genotypes for this gene, but some inbreeding occurs so that the frequency of identical homozygotes in the population is F, then the genotypic proportions are

$$\text{for } A_iA_i: \; Fp_i + (1 - F)p_i^2 = p_i^2 + Fp_i(1 - p_i)$$

$$\text{for } A_iA_j: \; 2(1 - F)p_ip_j = 2p_ip_j - 2Fp_ip_j \text{ for } i \neq j.$$

Thus, compared to the Hardy-Weinberg proportions, inbreeding produces an excess of homozygotes and this excess is proportional to the frequency of identical homozygotes in the population.

The frequency F of identical homozygotes is called the mean inbreeding coefficient of the population, or often simply the *population inbreeding coefficient*. If consanguinous matings are infrequent (as in Table 2-8), this population inbreeding coefficient can be calculated simply as the average coefficient of coancestry among mating pairs in the population of parents. For higher levels of consanguinity, the population inbreeding coefficient may be higher than this average.

Random genetic drift in a finite population in Hardy-Weinberg equilibrium has an effect analogous to inbreeding because the finite population size causes randomly chosen mates to have a nonzero coef-

Table 2-9. *Simple Coancestry Coefficients.* Coefficient of coancestry, f_{IJ}, between individuals I and J in some simple pedigrees.

Relationship	Pedigree	f_{IJ}
Sibs	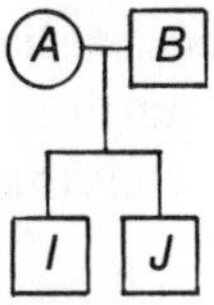	$\frac{1}{4}$
Half sibs	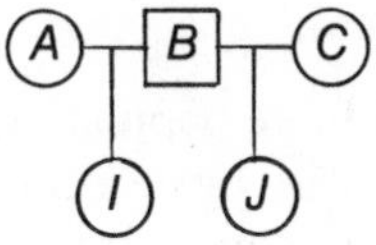	$\frac{1}{8}$
Uncle-niece	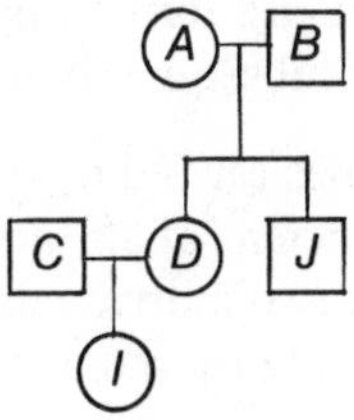	$\frac{1}{8}$
First cousins from sibs	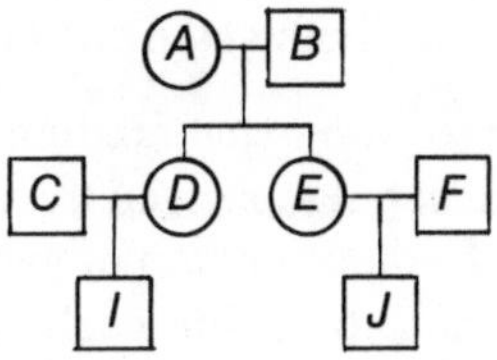	$\frac{1}{16}$
First cousins from half sibs	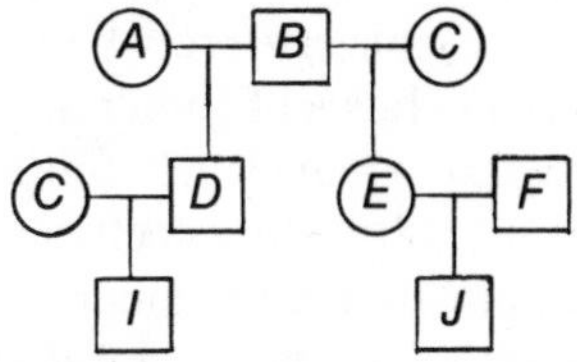	$\frac{1}{32}$

ficient of coancestry. The inbreeding coefficient of a random individual after t generations is

$$f_t = 1 - \left(1 - \frac{1}{2N}\right)^t (1 - f_0)$$

where N is the effective population size and f_0 is the inbreeding coefficient of the population in generation 0. Thus, as time goes on, f_t

becomes closer and closer to unity, when all individuals become homozygous. If all individuals in a randomly mating population are homozygous, the population must be monomorphic (in accordance with our previous results). In this sense inbreeding in small populations is due to the same stochastic process that produces the changes in allele frequencies. It is therefore not immediately comparable to the result for large populations with consanguineous matings where the gene frequencies are constant.

Exercise 2.4.B (Theoretical)

Random genetic drift in a finite population may be described as a process of progressive buildup of identity by descent. Imagine that N individuals shed a very large number of their gametes into a gamete pool, and that the zygotes of the offspring generation are formed by drawing two gametes at random from the gamete pool and subsequently uniting these gametes.

a. Assume that the $2N$ genes carried by the parents are all nonidentical, that is, the inbreeding coefficient of all individuals is 0 and the coefficient of coancestry between any two individuals is 0. Show that each offspring has probability $1/(2N)$ of being inbred at a given autosomal locus.
b. Under the assumption of (a), show that the coefficient of coancestry between two randomly chosen individuals in the offspring generation is $1/(2N)$.

The results in (a) and (b) show that random genetic drift builds up identity between the genes in the population even though the model describing reproduction is strictly one of random union of gametes. This buildup of identity has the property that the inbreeding coefficient of a random individual is equal to the coefficient of coancestry between two random individuals. The inbreeding is due to the nonzero probability that any gene is chosen more than once in the random choice of gametes. This is, of course, paralleled by a nonzero probability that a gene is not chosen at all.

Now suppose that the inbreeding coefficient with respect to a given autosomal gene in a random individual among the parents is f_o, and that the coefficient of coancestry of two random individuals is also f_o.

c. Show that the inbreeding coefficient of a random individual among the offspring is

$$f_1 = \frac{1}{2N} + \left(1 - \frac{1}{2N}\right)f_o.$$

d. Show that f_1 is the coefficient of coancestry between two random individuals in the offspring generation.
e. Now show that

$$f_t = \frac{1}{2N} + \left(1 - \frac{1}{2N}\right)f_{t-1}$$

where f_t is the probability of identity of two randomly chosen genes in the t'th generation (the inbreeding coefficient of a random individual or the coefficient of coancestry between random individuals) and f_{t-1} is the same probability among their parents.

f. Show

$$1 - f_t = \left(1 - \frac{1}{2N}\right)(1 - f_{t-1}),$$

and consequently that

$$1 - f_t = \left(1 - \frac{1}{2N}\right)^t (1 - f_o),$$

which is the expression given in section 2.4.1.

Exercise 2.4.C (Theoretical)

To illustrate the concept of effective population size (section 2.1.3), consider the model described in Exercise 2.4.B, but now assume that the population size is N_1 in odd numbered generations: $t = 2s - 1$, $s = 1,2,3,\ldots$, and N_2 in even numbered generations: $t = 2s$, $s = 0,1,2,\ldots$

a. Denote the probability of identity of two random genes in generation t by g_t and write g_{2s} as a function of the population size and the identity coefficient g_{2s-1} in the previous generation.
b. Write g_{2s-1} as a function of the population size and g_{2s-2} ($s = 1,2,\ldots$).
c. Combine the answers to a. and b. to form an expression for g_{2s} in terms of the population sizes and g_{2s-2}.

In Exercise 2.4.B we saw that

$$1 - f_t = \left(1 - \frac{1}{2N}\right)(1 - f_{t-1})$$
$$= \left(1 - \frac{1}{2N}\right)^2 (1 - f_{t-2}).$$

The *effective population size* when the population size varies is that value N_e such that

$$1 - g_{2s} = \left(1 - \frac{1}{2N_e}\right)^2 (1 - g_{2s-2}).$$

Thus, N_e is the population size that allows the inbreeding to increase at the same "average" rate as in the simple model of 2.4.B.

d. Find an expression for N_e.
e. Calculate N_e in the following cases:

1. $N_1 = 10$ $N_2 = 15$
2. $N_1 = 10$ $N_2 = 50$
3. $N_1 = 10$ $N_2 = 100$
4. $N_1 = 10$ $N_2 = 1000$

2.4.2 Assortative Mating

The preferential mating of like phenotypes, that is, *assortative mating*, has received much attention as a very plausible deviation from random mating (see sections 1.1.4.1 and 4.2). In the simplest model of genotypic assortative mating in a monogamous species we consider a locus where each genotype has a distinct phenotype, that is, we assume no dominance. A certain fraction h of the females choose a mate of the same phenotype (genotype), and the rest, a fraction $1 - h$, choose their mate at random. In this way all females are mated and if the genotypic proportions are equal between the sexes, then the genotypic proportions among breeding males are the same as among adult males, so no sexual selection occurs and no change in allele frequencies results. The genotypic proportions among zygotes, assuming no fecundity selection and no gametic selection, are given in Table 2-10. Thus the gene frequency is conserved. If no selection occurs, the genotypic proportions will converge to characteristic proportions given by

$$\text{for } A_iA_i : \hat{g}_{ii} = p_i^2 + Hp_i(1 - p_i) \text{ and}$$
$$\text{for } A_iA_j : \hat{g}_{ij} = 2(1 - H)p_ip_j,$$

where $H = (h/2)/(1 - h/2)$. These proportions are analogous to those obtained with inbreeding (if $H = F$). The convergence to these proportions is rather fast:

$$g'_{ii} - p_i^2 - Hp_i(1 - p_i) = (h/2)[g_{ii} - p_i^2 - Hp_i(1 - p_i)],$$
$$g'_{ij} - 2(1 - H)p_ip_j = (h/2)[g_{ij} - 2(1 - H)p_ip_j].$$

In the similar model of phenotypic assortative mating in a two-allele polymorphism with dominance, the assorting A – females choose A – males and the assorting aa females choose aa males. Here similar results

Table 2-10. *Mixed Assortative and Random Mating.*

Mate choice	Frequency	Offspring frequency	
		A_iA_i	A_iA_j
Assortative	h	$g_{ii} + \sum_{\substack{j=1 \\ j \neq i}}^{n} \frac{g_{ij}}{4}$	$\frac{g_{ij}}{2}$
Random	$1 - h$	p_i^2	$2p_ip_j$
Total		$g'_{ii} = \frac{h(p_i + g_{ii})}{2} + (1 - h)p_i^2$	$g'_{ij} = \frac{hg_{ij}}{2} + 2(1 - h)p_ip_j$

hold, but the characteristic genotypic proportions become considerably more complex.

Exercise 2.4.D (Theoretical)

Consider the assortative mating model of section 2.4.2 at an autosomal locus with two alleles A_1 and A_2. Let g_{11}, g_{12}, and g_{22} be the genotype frequencies and p_1, p_2 the allele frequencies.

a. Which of the female genotypes can produce A_1A_1 offspring when assorting?
b. Same question for A_2A_2 offspring.
c. Same question for A_1A_2 offspring.
d. Show that the genotypic frequencies among offspring from assorting females are $g_{11} + \frac{g_{12}}{4}, \frac{g_{12}}{2}$, and $g_{22} + \frac{g_{12}}{4}$.
e. Show that the genotypic frequencies in the offspring generation are

$$g_{11}' = h(p_1 + g_{11})/2 + (1 - h)p_1^2$$

$$g_{12}' = hg_{12}/2 + 2(1 - h)p_1p_2$$

$$g_{22}' = h(p_2 + g_{22})/2 + (1 - h)p_2^2.$$

f. Show that $g_{11}' = g_{11}$, $g_{12}' = g_{12}$ and $g_{22}' = g_{22}$ when

$$g_{11} = \hat{g}_{11} = p_1^2 + Hp_1p_2$$

$$g_{12} = \hat{g}_{12} = 2(1 - H)p_1p_2$$

$$g_{22} = \hat{g}_{22} = p_2^2 + Hp_1p_2,$$

where $H = (h/2)/(1 - h/2)$.

g. Show that

$$g_{11}' - \hat{g}_{11} = (h/2)(g_{11} - \hat{g}_{11})$$

$$g_{12}' - \hat{g}_{12} = (h/2)(g_{12} - \hat{g}_{12})$$

$$g_{22}' - \hat{g}_{22} = (h/2)(g_{22} - \hat{g}_{22}),$$

and use this to argue that the genotypic proportions in the population converge very quickly to $\hat{g}_{11}$, $\hat{g}_{12}$, and $\hat{g}_{22}$.

h. Now show that the same results hold for multiple alleles (section 2.4.2 and Table 2-10).

2.4.3 Geographically Structured Populations

Human populations are rarely the ideal, homogeneous randomly mating isolates we have considered in previous sections. Such populations are called *panmictic* to distinguish them from populations that have geographical structure in the sense that closely situated individuals have a higher probability of mating than those far apart. The primary effect of geographical structure is therefore to produce deviations from random mating and to produce genotypic proportions that may deviate from the Hardy-Weinberg proportions.

The simplest model of a partitioned population is the island model. Here the population consists of a number n of isolated islands with N_i, $i = 1, 2, \ldots, n$, inhabitants. The total population size is then

$$N_O = \sum_{i=1}^{n} N_i,$$

and the relative sizes of the islands are $c_i = N_i/N_O$, $i = 1, 2, \ldots, n$. Let p_i be the frequency of an autosomal allele A on the ith island, then the frequency of allele A in the total population is

$$p_O = \sum_{i=1}^{n} 2N_i p_i/(2N_O) = \sum_{i=1}^{n} c_i p_i.$$

Suppose mating is random on each island so that Hardy-Weinberg proportions apply to zygotes, then the genotypic frequency of AA in the total population is

$$\sum_{i=1}^{n} c_i p_i^2.$$

This frequency deviates from the Hardy-Weinberg proportions corresponding to the gene frequency p_O in the total population by

$$V = \sum_{i=1}^{n} c_i p_i^2 - p_O^2 = \sum_{i=1}^{n} c_i (p_i - p_O)^2 > 0.$$

This deviation is exactly the variance in allele frequency among the islands, and we see that the number of homozygotes in the *total* population is always larger than expected from the Hardy-Weinberg proportions in that population. This result is known as the *Wahlund principle.* For a two-allele polymorphism, the genotypic proportions in the total population are

$$\begin{aligned} &\text{for } AA\text{: } p_O^2 + V, \\ &\text{for } Aa\text{: } 2p_O q_O - 2V, \text{ and} \\ &\text{for } aa\text{: } q_O^2 + V. \end{aligned}$$

(See Table 2-11.)

These proportions are similar to those for inbreeding with the population inbreeding coefficient $F = V/(p_O q_O)$. For multiple alleles, the parallel to inbreeding breaks down because the variance in allele frequency may vary among alleles. The change in heterozygote frequency is twice the covariance among populations in the frequency of the alleles in the heterozygote, and this may be positive or negative.

Exercise 2.4.E

Consider the data of Table 2-11.

a. Calculate the gene frequencies in the four populations and samples.
b. In what proportions should fish from populations 1 and 2 be mixed to give the gene frequencies observed in samples 1 and 2?
c. Calculate the variance in gene frequencies of samples 1 and 2 using the mixing proportions from (b).
d. Calculate the expected genotypic proportions in samples 1 and 2 under the simple assumption that the fish in the samples are mixtures of fish from populations 1 and 2.

e. Is this simple assumption a reasonable explanation of the observed genotypic proportions?

Such genotypic proportions occur in a population that is formed by the mixture of immigrants from an array of populations with varying allele frequencies. Thus a newly mixed population is characterized by an excess of homozygotes. After one generation of panmictic random mating, however, the Hardy-Weinberg proportions are restored and the mixture forgotten.

Migration among the islands may alter the gene frequency of any island, but if the emigrations and immigrations do not alter fitness of the individuals on an island, then the total allele frequency is preserved; the migration will have a smoothing effect on the variation in gene frequency. If all populations interchange migrants, then after a sufficiently long time they will all have an allele frequency p_O.

Random genetic drift causes isolated populations to diverge, so for finite populations in an island model there are two opposing forces:

Table 2-11. *Genotypic Proportions in a Mixed Population.* The cod, *Gadus morhua,* is polymorphic for the alleles A_1 and A_2 at a hemoglobin locus. The populations within Danish waters are rather homogeneous, with a frequency of allele A_1 of about 0.6. The allele A_1 is considerably rarer in the Baltic Sea. In an area in the western part of the Baltic Sea the population is rather heterogeneous and the frequency may change in time. Samples 1 and 2 are taken geographically close together in this area. The Hardy-Weinberg proportions corresponding to the observed allele frequencies are given for each population and each sample in terms of the expected genotypic distributions (in italics) in the sample based on Hardy-Weinberg proportions.

	A_1A_1	A_1A_2	A_2A_2	Total
Population 1	258	324	106	688
(Southeast Denmark)	*256.4*	*327.2*	*104.4*	*688*
Population 2	0	10	230	240
(Northern Baltic)	*0.1*	*9.8*	*230.1*	*240*
Sample 1	67	118	81	266
(Western Baltic)	*59.7*	*132.6*	*73.7*	*266*
Sample 2	49	73	128	250
(Western Baltic)	*29.2*	*112.6*	*108.2*	*250*

Data from Sick (1965).

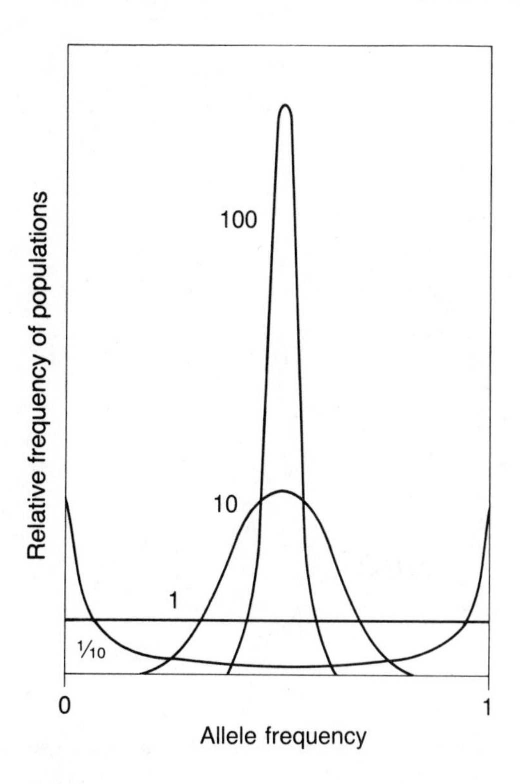

Figure 2-5. The distribution of allele frequencies among subpopulations in Wright's island model with an average allele frequency of 0.5. (After Wright, 1951.) The numbers are values of 2Nm *used to provide the curves.*

the smoothing force of migration and the diverging force of random genetic drift. The balance between these two forces depends on the specific migration structure assumed for the model. The simplest structure is that of Wright's island model, where all populations are of the same size ($N_i = N$) and all supply the same number of emigrants. The emigrants are mixed and returned equally to all populations. Any two populations are at the same distance as any other two, and in this sense the model lacks geographical structure. Here the interaction of migration and drift is very simple. The populations diverge as if they were isolated, provided each population supplies and receives less than one migrant per generation; the smoothing effect of migration predominates if migration rates per population per generation are higher (Figure 2-5). This result holds when the number of

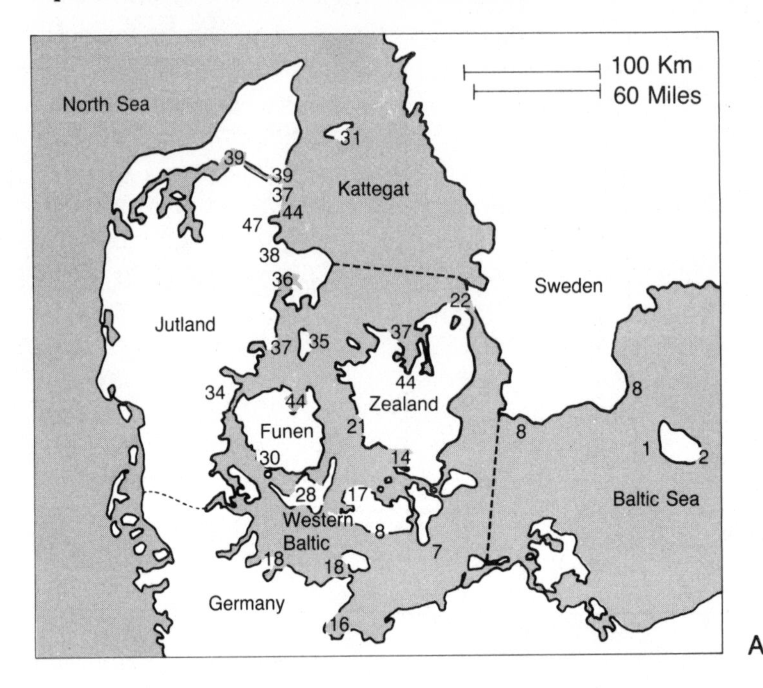
100 Km
60 Miles
North Sea
Kattegat
Sweden
Jutland
Zealand
Funen
Western
Baltic
Baltic Sea
Germany
31
39
39
37
44
47
38
36
22
37
35
37
44
34
44
21
8
8
1
2
30
14
28
17
8
7
18
18
16
A

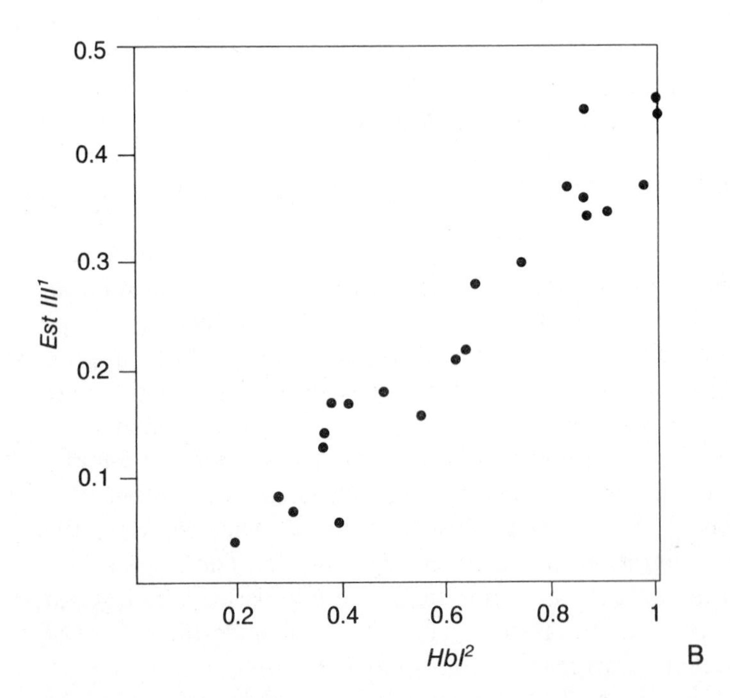
0.5
0.4
0.3
0.2
0.1
$Est\ III^1$
0.2
0.4
0.6
0.8
1
Hbl^2
B

subpopulations is so large that random genetic drift can be neglected in the total population.

A model with more geographical structure is the steppingstone model of Kimura and Weiss. Here the islands are ordered in a long chain and migration only occurs between neighboring populations. The results of this subdivision are not straightforward because, even with rather high migration between neighboring populations, two populations far apart in the chain may be virtually isolated from each other. With very low migration rates the model degenerates to Wright's island model with low migration and all populations diverge. For higher migration rates, populations sufficiently far apart will diverge due to genetic drift, but locally the smoothing effect of migration takes over. This makes the gene frequency vary smoothly between extremes over the population chain and the local migration rate determines the steepness of the allele frequency variation (see Kimura and Ohta, 1971). This model therefore produces clines in allele frequency, that is, areas where the gene frequency varies monotonically from one level to another as, for example, the cline in the frequency of I^B of the *ABO* system between Asia and Europe or the clines in Figure 2-6.

Geographical variation in allele frequencies at two loci produces new phenomena. In our island model with isolated populations, let the frequency of allele A be p_{A_i} in the ith population and that of allele B of another locus be p_{B_i}. Further, let x_i be the frequency of gamete AB AB in the ith population and c the recombination frequency. Then if all populations are mixed, the linkage disequilibrium in the offspring population becomes

$$(1 - c)\overline{D} + C$$

Figure 2-6 (opposite). A: *Geographical variation in the frequency of the* EstIII^1 *allele in* Zoarces viviparus. *The allele frequency varies in a cline from about 0.4 in Kattegat to about 0.05 in the Baltic Sea. Across the same geographical area, and at the same time, the* HbI^2 *allele varies from a frequency close to 1 in Kattegat to about 0.1 in the Baltic Sea.* B: *Parallel variation in the frequency of the* EstIII^1 *and* HbI^2 *alleles in the area limited by the broken lines in* A. *This linear relationship is expected if the cline is due to migration between neighboring populations in the area between Kattegat and the Baltic Sea (Christiansen and Frydenberg, 1974). (Data from Frydenberg et al., 1973 and Hjorth and Simonsen, 1975. After Christiansen, 1977.)*

where

$$\overline{D} = \sum_{i=1}^{n} c_i(x_i - p_{A_i}p_{B_i})$$

is the mean linkage disequilibrium among the parents,

$$C = \sum_{i=1}^{n} c_i(p_{A_i} - p_{A_0})(p_{B_i} - p_{B0})$$

is the covariance in allele frequency at the two loci, and

$$p_{A_0} = \sum_{i=1}^{n} c_i p_{A_i}$$

and

$$p_{B_0} = \sum_{i=1}^{n} c_i p_{B_i}$$

are the mean allele frequencies. Thus the mixture of the populations creates a linkage disequilibrium among offspring equal to the covariance in allele frequencies between the two loci (we cannot define the total linkage disequilibrium in the population before mixing because, in our definition of linkage disequilibrium in section 2.3, we assumed random mating). This is the two-locus equivalent of the Wahlund principle for population mixing.

In a geographically structured population the creation of linkage disequilibrium by migration, combined with the finite rate of decay, can build up rather large levels of linkage disequilibrium in the local populations. The linkage disequilibrium in a local population settles on a value that is proportional to the covariance in allele frequencies at the two loci in the mixture after immigration to the local population, and inversely proportional to the frequency of recombination (Feldman and Christiansen, 1975).

Exercise 2.4.F (Theoretical)

In the island model of section 2.4.3, consider two loci with alleles A,a and B,b. Let x_i, p_{Ai}, and p_{Bi} be the frequencies of gamete AB, of allele A, and of allele B, respectively, in the ith population.

a. Show that if all the populations are mixed, then the frequency of the *AB* gamete in the next generation is

$$xi' = \sum_{i=1}^{n} c_i[(1 - c)x_i + cp_{A_i}p_{B_i}].$$

b. Show that this implies that the linkage disequilibrium $D' = x' - p_{A_O}p_{B_O}$ among the offspring is given by

$$D' = (1 - c)\overline{D} + C,$$

where

$$C = \sum_{i=1}^{n} c_i(p_{A_i} - p_{A_O})(p_{B_i} - p_{B_O}).$$

(Hint: write $x_i = p_{A_i}p_{B_i} + D_i$ where D_i is the linkage disequilibrium in the *i*'th population.)

3

Changes in Allele Frequencies

In most branches of science it is departures from simple rules that initiate progress. In this sense it is important to determine and classify the biological phenomena that produce departures from the basic population genetic rules outlined in Chapter 2. Mutation, the original source of genetic variation, produces deviations from the strict assumptions of Mendel's first law, and therefore from the law of allele frequency conservation. In local populations immigration can change the allele frequencies (section 2.4.3) if geographic variation exists. For the global population, however, the allele frequencies are conserved. Thus, even though migration is locally a source of genetic variance, it is fundamentally different from mutation.

The other major cause of change in the genetic composition of a population over time is natural selection, which occurs whenever there is variation among genotypes in the contribution that individuals make to future generations. In this chapter we describe the effect of these evolutionary forces, mutation and natural selection, on the simple genetic variation considered in Chapter 2. We address in detail only the simpler forms of natural selection although we allude to some of the more complex ideas near the end of the chapter.

3.1 Mutation

In the arguments leading to the law of allele frequency conservation, we made an admittedly unrealistic assumption, namely, that the transmission of the genetic material is completely conservative. Genes mutate, but the frequency of mutation is very low (typically less than 10^{-4} per generation), so the change in gene frequency possibly mediated by mutation is indeed very small. In terms of allele frequency conservation over long time spans, however, mutation becomes relevant.

3.1.1 Two-allele Mutation Balance

To understand how mutation can affect populations, consider a very simple model of reversible mutation between two autosomal alleles A_1 and A_2. Assume that A_1 mutates to A_2 with the frequency μ_1 per generation and that A_2 mutates to A_1 with frequency μ_2. Then the allele frequencies, p_1 and p_2, change to

$$p_1' = (1 - \mu_1)p_1 + \mu_2 p_2 \text{ and } p_2' = (1 - \mu_2)p_2 + \mu_1 p_1,$$

respectively, in the following generation. The first equation can be modified to

$$p_1' = (1 - \mu_1 - \mu_2)p_1 + \mu_2,$$

and this change in allele frequency from generation to generation can be graphed as in Figure 3-1. For small initial p_1 the frequency of A_1 will increase ($p_1' > p_1$), and for large p_1 the frequency will decrease ($p_1' < p_1$). At the intermediate allele frequency,

$$\hat{p}_1 = \frac{\mu_2}{\mu_1 + \mu_2},$$

no change in the allele frequency is expected and the population is at equilibrium with respect to mutation. In populations with $p_1 < \hat{p}_1$ initially, the frequency of A_1 will gradually increase until $\hat{p}_1$ is reached and vice versa for $p_1 > \hat{p}_1$, since

$$p_1' - \hat{p}_1 = (1 - \mu_1 - \mu_2)(p_1 - \hat{p}_1).$$

Thus after n generations we have

$$p_1^{(n)} - \hat{p}_1 = [1 - (\mu_1 + \mu_2)]^n(p_1 - \hat{p}_1).$$

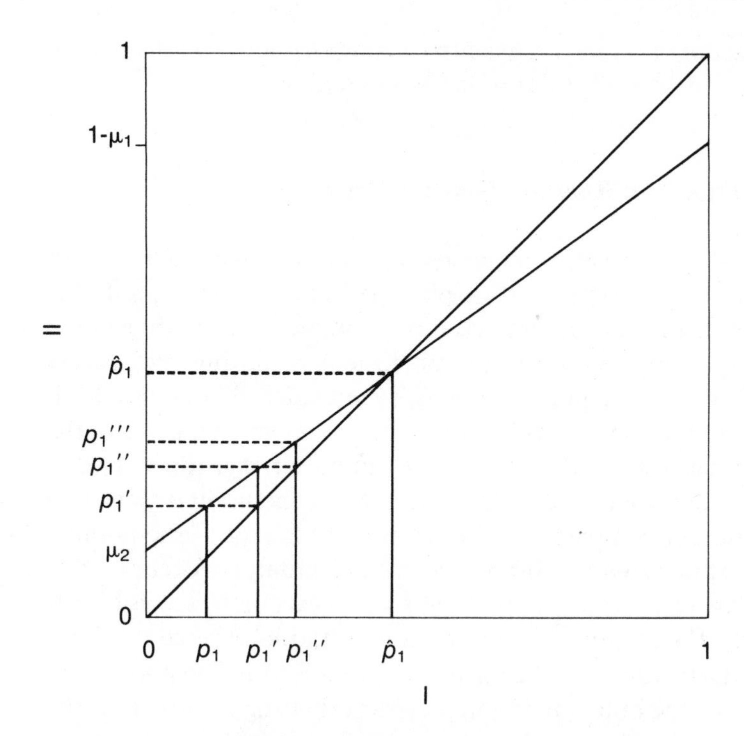

Figure 3-1. The change in allele frequency from generation to generation due to mutation in the model of section 3.1.1. Corresponding to the allele frequency in the parent population on axis I, the allele frequency in the next generation is given on axis II. The convergence to the globally stable equilibrium $\hat{p}_1$ *is indicated. Mutation rates shown here are unrealistically large.*

Consequently, after a sufficiently long time, the deviation in allele frequency from the equilibrium $\hat{p}_1$ is very small. We say that the allele frequency equilibrium $\hat{p}_1$ is *stable,* and, because convergence to this equilibrium is guaranteed from any initial state, the equilibrium is said to be *globally stable*. The approach to this equilibrium, however, is extremely slow. The number of generations required to come halfway to equilibrium from any initial allele frequency is approximately

$$n_{1/2} \approx 0.7/(\mu_1 + \mu_2),$$

which for large mutation frequencies of the order of 10^{-4} is about 3500 generations. Thus convergence to this mutation-mutation equilibrium is only interesting on an evolutionary time scale, and by comparison with the effect of mutation on gene frequency change, random genetic drift becomes important, even in large populations.

3.1.2 Mutation and Random Genetic Drift

Processes as slow as genetic change under mutation may interact with genetic drift because the time scale of allele frequency changes due to the two phenomena may be similar. First, however, it should be noted that mutation is important for our conclusions regarding the process of genetic drift in finite populations. We concluded that after a sufficiently long time any population will become monomorphic due to the stochastic variations in allele frequency, and after that no change in the allele frequency will occur. Obviously this conclusion depends on the assumption of completely conservative transmission; the population is not trapped in a monomorphic state if mutations occur.

The process of divergence in allele frequency due to genetic drift interacts with the attainment of mutation-mutation balance in a very simple way. Depending on the *number* of mutations in the population per generation, the outcome in most cases is that the results from the theory of random genetic drift (section 2.1.3, if we neglect mutation) or the results from the mutation theory (section 3.1.1, if we neglect random drift) hold approximately. The number of mutations per generation in the population is $2N\mu$, where N is the effective population size and μ is the average mutation rate per gene per generation. If the number of mutations is large, $2N\mu >> 1$, then we can neglect genetic drift, and if this number is small, $2N\mu << 1$, then mutation can be neglected in that the changes in allele frequency will be mediated by drift except when the population becomes monomorphic and new mutations occur with long time intervals between them. For $2N\mu \approx 1$, an intermediate situation occurs where the allele frequency is expected to show large fluctuations about the mutation-mutation balance equilibrium.

3.1.3 Evolution by Neutral Substitutions

These results for the interaction between mutation and drift hold for more realistic one locus multiple allele models as long as the assumption of no selection is maintained (see Kimura and Ohta, 1971; Ewens,

1979; and Kimura, 1983). For a set of alleles, $A_1, A_2, \ldots, A_k$, which are selectively neutral (that is, natural selection does not discriminate between genotypes formed by these alleles), the processes of random genetic drift and mutation can change the appearance of the population through time.

Suppose the population is initially monomorphic for allele A_1, and that single mutation events occur in the population (i.e., $2N\mu$ is less than about 1). Each new mutant is initially at an allele frequency of $1/(2N)$, and from the law of allele frequency conservation a new mutant has the probability $1 - 1/(2N)$ of being lost from the population and $1/(2N)$ of increase to fixation. Although the probability of fixation of a new allele is very small in a large population, given enough time some of the many new mutants are bound to take over the population. The number of mutants per generation is $2N\mu$ and their probability of fixation is $1/(2N)$, so that on the average μ fixations of a new allele occur per generation. Thus, we expect that after $1/\mu$ generations, our population, initially monomorphic for A_1, will have another allele, say, A_2, substituted for A_1. Evolution will therefore occur at this locus with an average allele substitution rate equal to the mutation rate independent of the population size. This process is known as *evolution by neutral allele substitutions*.

The average time between neutral substitutions is equal to the inverse of the mutation rate to neutral alleles. This time interval, however, is exactly the time needed for each gene to have mutated an average of one time. Thus, if the number of possible alleles is large enough that mutation back to previously existing alleles can be neglected, then we do not have to assume that the population is finite to produce the result. Each gene has on the average mutated to a new allele in the same span of $1/\mu$ generations, and on the average all alleles present disappear after $1/\mu$ generations.

This constant rate of substitution for neutral alleles has led to their consideration as a *molecular clock* over long time spans. Consider the locus for the protein cytochrome-C and suppose that the amino acid sequence of this protein in one of the early ancestors of mammals were known. Now count as mutations any genetic change that alters one of the protein's amino acids; then we can count the number of substitutions necessary to explain the change from the ancestral protein to cytochrome-C in, say, humans. Although the amino acid sequence of the old gene is not observable, we can use the molecular clock idea to calculate the minimum number of allele substitutions that separate humans and horses from their common ancestor. This is to be interpreted as the sum of the number of substitutions in the line leading to humans and the line leading to horse. This sum is called the *genetic distance* between human and horse with respect to cytochrome-C. The information in these genetic distances among a group of related species can be used to construct a tree showing the relationship among the species (Figure 3-2). Usually the relationship among species described in this tree coincides rather closely with the relationships among the species derived from other criteria. For example, the genetic distances back to common ancestors are roughly proportional to the time inferred from the fossil record. This proportionality allows estimation of the number of allele substitutions per time unit. For cytochrome-C the estimate is about 0.3 amino acid substitutions per amino acid per 10^9 years, a rather low figure (Table 3-1).

The pattern of cytochrome-C observations is compatible with the hypothesis that the majority of substitutions have been neutral and, if the mutation rate to neutral alleles is constant over time, the observations are compatible with our picture of a process of evolution by neutral substitutions (Kimura, 1983).

Population size is important in determining the population's genotypic composition while undergoing neutral evolution. If the number of neutral alleles at the considered locus is very large, then the frequency of heterozygotes in the population, the *heterozygosity*, is about $H = 4N\mu/(1 + 4N\mu)$ when averaged over a sufficiently long time. When $2N\mu$ is small and drift is important, the average heterozygosity

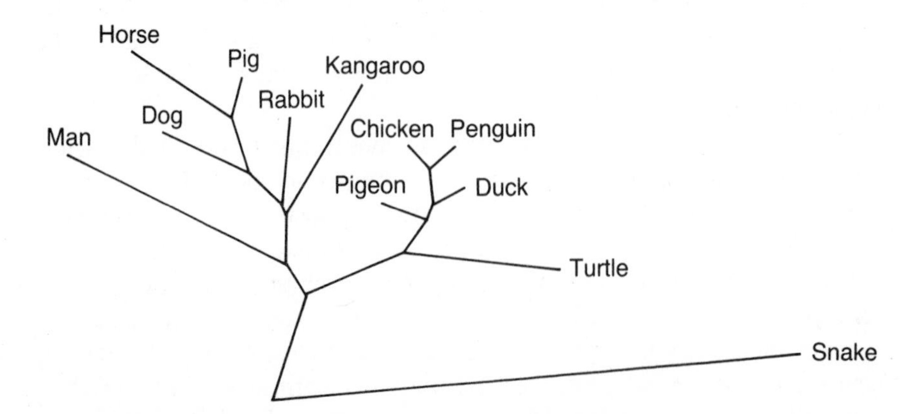

Figure 3-2. Evolutionary tree constructed from amino acid sequence comparisons between species. (After Fitch and Margoliash, 1967.)

Table 3-1. *Rates of Substitutions in Mammals.* Substitution rates of amino acids as estimated by comparison of the amino acid sequences among mammalian species.

Protein	$10^9 \times$ substitutions per amino acid per year
Cytochrome-C	0.3
Insulin	0.4
Hemoglobin α-chain	1.2
Pancreatic ribonuclease	2.1
Fibrinopeptides	8.3

After Kimura (1983).

is small. Since substitutions do occur, however, such a low average heterozygosity must indicate that although the population is usually monomorphic, there are rare episodes of polymorphism during the process of substitution. When $2N\mu$ is larger, the heterozygosity takes larger values; for example, for $2N\mu > 1/2$ we have $H > 1/2$, in which case the population is usually polymorphic for three or more alleles (for two-allele polymorphisms, $H < 1/2$). Thus for large values of $2N\mu$ the population will usually be polymorphic for many alleles, and evolution by neutral allele substitutions occurs as one set of polymorphic alleles is replaced by another set.

For humans the average heterozygosity for electrophoretic loci is 6–7 percent (section 1.2.5) and for fruit flies it may be as high as 18 percent (Table 1-12). These average figures are difficult to relate to this theory, however, because the mutation rate may vary among the loci studied (Table 3-1; Koehn and Eanes, 1978).

Exercise 3.1.A (Theoretical)

Show that the frequency of heterozygotes in a two-allele autosomal polymorphism is always less than 1/2 when the genotypic frequencies are in Hardy-Weinberg proportions.

Exercise 3.1.B (Theoretical)

Suppose the population is initially monomorphic for allele A_1, and suppose further that all the genes in the population are identical by descent, that is, $f_o = 1$ (cf. Exercise 2.4.B). Now assume that any mutation is to a new allele.

a. Show that the probability of identity of two randomly chosen genes (cf. Exercise 2.4.B) is

$$f_t = (1-\mu)^2\left[\frac{1}{2N} + \left(1 - \frac{1}{2N}\right)f_{t-1}\right],$$

where μ is the mutation rate to new alleles.

b. Show that this identity coefficient, f_t, converges to

$$f_\infty = \frac{(1-\mu)^2}{2N - (1-\mu)^2(2N-1)}.$$

c. Show that the identity coefficient f is the frequency of homozygotes in the population, and that $1 - f$ is the frequency of heterozygotes, H, in the population.

Since μ is usually a very small number, the approximations

$$f_\infty \approx \frac{1}{1 + 4\mu N},$$

and

$$H = 1 - f_\infty = \frac{4N\mu}{1 + 4N\mu}$$

may be used in applications.

3.2 Selection

Much of the single gene variation in human populations reviewed in section 1.2 involves alleles causing diseases or alleles showing elevated frequencies among individuals with certain diseases. These diseases may alter an individual's chance of survival, they may alter the probability of reproducing, or they may influence the number of children. All of these provide the opportunity for natural selection to work. We will illustrate the changes in allele frequencies mediated by selection by considering some very simple models of natural selection and then discuss these models in terms of the aforementioned diseases.

3.2.1 Zygotic Selection of One Gene

Consider two alleles A_1 and A_2 and suppose that the probability of survival from zygotes to adults is v_{11}, v_{12}, and v_{22} for the three genotypes A_1A_1, A_1A_2, and A_2A_2, respectively. No sexual, gametic, or fecundity selection occurs. Suppose further that reproduction occurs by random mating so that the genotypic frequencies among newly formed zygotes are p_1^2, $2p_1p_2$, and p_2^2, the Hardy-Weinberg proportions. After zygotic selection, the genotypic frequencies will be $p_1^2v_{11}/w$, $2p_1p_2v_{12}/w$, and $p_2^2v_{22}/w$, where $w = p_1^2v_{11} + 2p_1p_2v_{12} + p_2^2v_{22}$. Thus, the allele frequency of A_1 among adults is

$$p_1' = \frac{p_1^2v_{11} + p_1p_2v_{12}}{w},$$

which is also the allele frequency among zygotes in the following generation. We can rewrite this equation as

$$p_1' = \frac{p_1w_1}{w},$$

where $w_1 = v_{11}p_1 + v_{12}p_2$ is the mean probability of survival of allele A_1 from zygotes to adults (since a fraction p_1 of the A_1 alleles in the population is in genotype A_1A_1 and a fraction p_2 is in A_1A_2), with $w_2 = v_{12}p_1 + v_{22}p_2$ the analogous probability for A_2. Here, $w = p_1w_1 + p_2w_2$ can be understood as the mean survival of individuals or alleles in the population. Hence the change in frequency of a given allele is determined by the ratio of the mean viability of that allele relative to the mean viability of the population.

The viability of an allele changes with the allele frequencies. When A_1 is rare and p_1 is close to 0, then we have $w_1 \approx v_{12}$, and $w \approx v_{22}$, so $p_1' > p_1$ when $v_{12} > v_{22}$ and $p_1' < p_1$ when $v_{12} < v_{22}$. There are four different cases of change in rare allele frequencies, as shown in Table 3-2.

Table 3-2. *Cases of Selection on a Single Locus.*

Selection	A_1 rare	A_2 rare
1. $v_{11} > v_{12} > v_{22}$	A_1 increases	A_2 decreases
2. $v_{11} < v_{12} < v_{22}$	A_1 decreases	A_2 increases
3. $v_{11} > v_{12}$, $v_{12} < v_{22}$	A_1 decreases	A_2 decreases
4. $v_{11} < v_{12}$, $v_{12} > v_{22}$	A_1 increases	A_2 increases

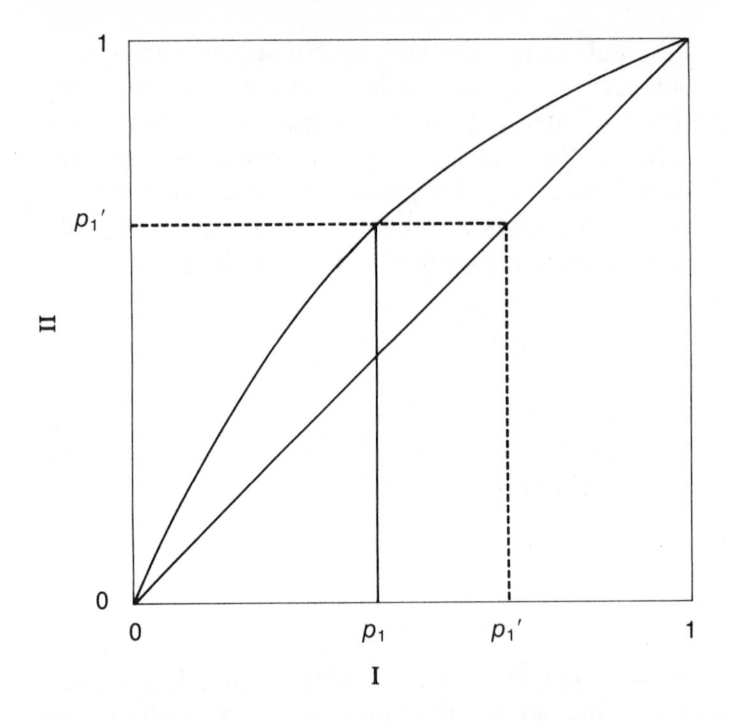

Figure 3-3. The change in allele frequency between generations in selection case 1, directional selection for allele A_1. *Corresponding to the allele frequency of the parent population* p_1 *on axis I, the allele frequency in the offspring generation* p_1' *is given on axis II.*

In case 1, $w_1 > w_2$ for all allele frequencies, so $w_1 \geqslant w$. Thus $p_1' > p_1$ for all initial allele frequencies p_1 ($0 < p_1 < 1$, Figure 3-3), and the allele frequency of A_1 will increase every generation and approach the *globally stable* equilibrium $p_1 = 1$. At this equilibrium allele A_1 is fixed in the population, which is therefore monomorphic A_1A_1. This case is known as *directional selection* for allele A_1. Case 2 is the analogous situation of directional selection for allele A_2 (Figure 3-4).

Cases 3 and 4 are more complicated in that the direction of change is opposite for p_1 close to 0 and 1. For both cases there exists a gene frequency $\hat{p}_1$, where $w_1 = w_2$, at which no change in allele frequency occurs, that is, $p_1' = p_1$ for $p_1 = \hat{p}_1$. The population is at a polymorphic allele frequency equilibrium when $p_1 = \hat{p}_1$, with

$$\hat{p}_1 = \frac{v_{12} - v_{22}}{2v_{12} - v_{11} - v_{22}}.$$

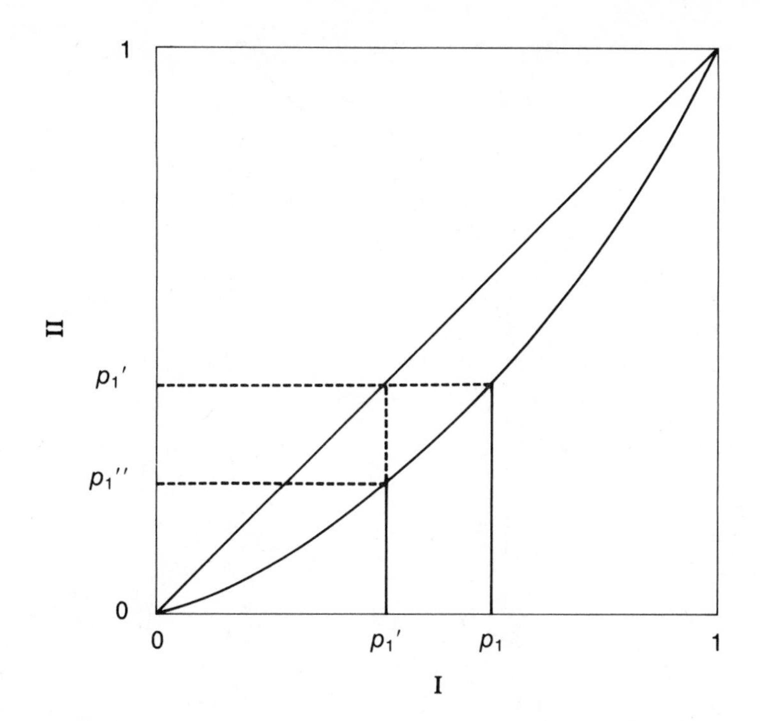

Figure 3-4. The change in allele frequency between generations in selection case 2, directional selection for allele A_2. *Axes are as in Figure 3-3. The convergence to the globally stable equilibrium* $\hat{p}_1 = 0$ *is indicated.*

In case 3, *underdominant selection,* $\hat{p}_1' < p_1$ for $p_1 < p_1'$ and $p_1' > p_1$ for $p_1 > \hat{p}_1$ (Figure 3-5). Thus, if the allele frequency differs from $\hat{p}_1$, then it will move away from $\hat{p}_1$ in the following generations; the equilibrium is said to be *unstable*. In this case, if initially $p_1 > \hat{p}_1$, then ultimately A_1 approaches fixation in the population. This equilibrium is not globally stable as it is only approached if initially $p_1 > \hat{p}_1$. To underscore this, it is said to be *locally stable*. Thus in case 3 the population will ultimately reach one of the locally stable monomorphic equilibria A_1A_1 or A_2A_2 depending on the initial allele frequencies.

In selection cases 1, 2, and 3 one of the alleles is ultimately lost from the population and only in case 4 is the maintenance of polymorphism possible. In case 4, *overdominant selection* (Figure 3-6), the polymorphic equilibrium, $\hat{p}_1$, is globally stable.

For multiple alleles the same method applies to the analysis of zygotic selection. The frequency of allele A_i changes according to

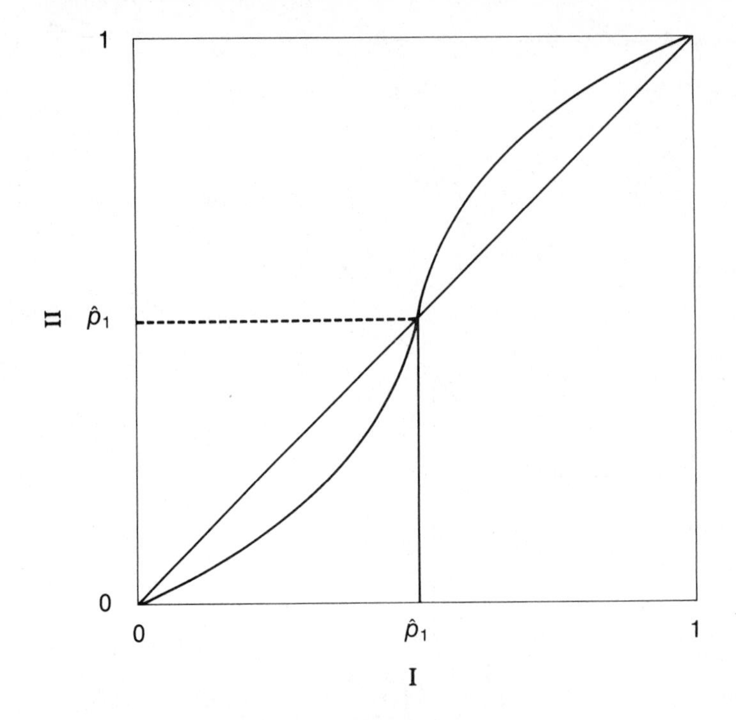

Figure 3-5. The change in allele frequency between generations in selection case 3, underdominant selection. Axes are as in Figure 3-3. The unstable equilibrium $\hat{p}_1$ *is shown.*

$$p_i' = \frac{p_i w_i}{w},$$

where

$$w_i = \sum_j v_{ij} p_j, \quad w = \sum_j w_j p_j,$$

and v_{ij} is the viability of genotype A_iA_j. Although the number of possible equilibria of the population is greatly increased, there can never be more than a single completely polymorphic equilibrium (i.e., one for which all $\hat{p}_i > 0$). At all of the other equilibria one or more alleles are missing.

Since the changes in allele frequencies are determined by the relative viabilities w_i/w, the absolute values of genotypic viabilities are

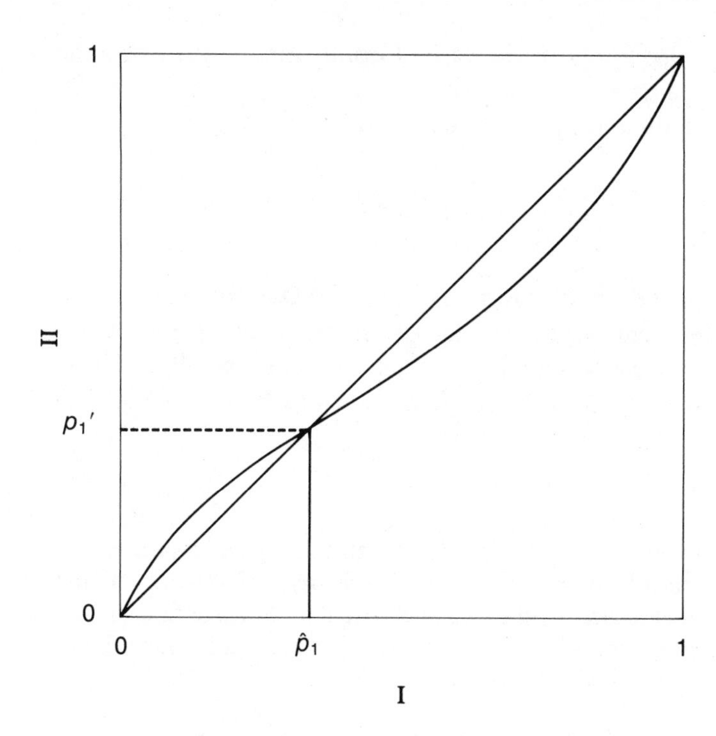

Figure 3-6. The change in allele frequency between generations in selection case 4, overdominant selection. Axes are as in Figure 3-3. The globally stable equilibrium $\hat{p}_1$ *is shown.*

unimportant in that a multiplication of the relative values by a constant leaves the results unaltered. For this reason, in theoretical studies the viabilities are often replaced by their values relative to that of a given genotype. For example, in the two-allele case the viabilities v_{11}, v_{12}, v_{22} can be replaced by the fitnesses relative to the homozygote A_1A_1, that is, 1, v_{12}/v_{11}, v_{22}/v_{11}, or those relative to the heterozygote A_1A_2, that is v_{11}/v_{12}, 1, v_{22}/v_{12}. Viability differences are often expressed in terms of differences from the viability of the normalized genotype, using *selection coefficients*. When normalized to A_1A_1 these are

$$s_{11} = \frac{v_{11}}{v_{11}} - 1 = 0, \ s_{12} = \frac{v_{12}}{v_{11}} - 1, \text{ and } s_{22} = \frac{v_{22}}{v_{11}} - 1,$$

in which case the relative viabilities become 1, $1 - s_{12}$, $1 - s_{22}$.

Cases 1 and 2 also apply in the case of dominance, e.g., when the relative viability of aa is $1 - s$ relative to that of the dominant phenotype A−. If the allele frequencies of A and a are p and q, then

$$p' = \frac{p}{1 - sq^2} \text{ and } q' = \frac{q(1 - sq)}{1 - sq^2}.$$

Thus when $s > 0$, then $p' > p$ and $q' < q$ for all allele frequencies and the population will end up at the globally stable monomorphic equilibrium AA. The *carbonaria* allele in *Biston betularia* is an allele whose increase can be described in this way, as shown in Figure 3-7.

Exercise 3.2.A

The moth *Panaxia dominula* is polymorphic with three phenotypes, *typica*, *medionigra*, and *bimacula*, corresponding to the genotypes D_1D_1, D_1D_2, and D_2D_2 at an autosomal locus. In an experiment equal numbers of females and males of the phenotype *medionigra* were allowed to mate and lay eggs. Later the following adults hatched from the pupae:

Phenotype:	*Typica*	*Medionigra*	*Bimacula*
Number:	36	42	21

From Sheppard and Cook, 1962.

a. Assuming Mendelian segregation of the alleles D_1 and D_2 calculate the relative viabilities of the three genotypes.
b. What will happen to the allele frequencies in this experimental population if only zygotic selection is working on the polymorphism?

Exercise 3.2.B

Drosophila pseudoobscura is polymorphic for gene arrangements (inversions) on its third chromosome. This polymorphism has been extensively studied by Dobzhansky and colleagues, and the following experiment is from Dobzhansky and Levene (1951).

The egg-to-adult viabilities of the standard gene arrangement (ST), the Chiricahua (CH) gene arrangement, and the heterokaryotype ST/CH have been studied. The gene arrangements are stable and they may formally be treated as alleles at a single locus. In *Drosophila* a karyotype like ST/CH has normal fertility and shows Mendelian segregation of the two arrangements.

A large number of flies of karyotype ST/CH mated and laid their eggs in a population cage. A sample of the offspring exhibited the following composition:

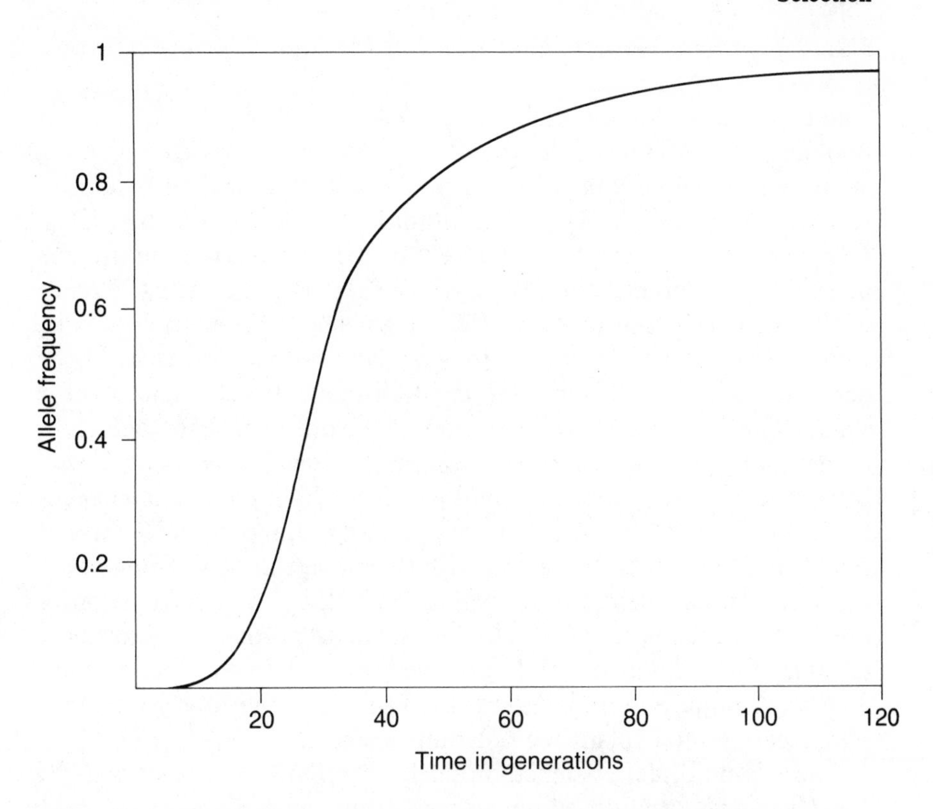

Figure 3-7. Increase in the frequency of the carbonaria *allele in* Biston betularia, *assuming a viability of 1.3 of* carbonaria *relative to the viability of the* typica *morph. The allele frequency at generation 0 is 0.001.*

	CH/CH	ST/CH	ST/ST	Total
Females	34	117	49	200
Males	35	117	48	200
Total	69	234	97	400

a. Calculate the relative viabilities of the three karyotypes.

b. Predict the evolutionary fate of this experimental population.

Exercise 3.2.C

In humans and other mammals heterokaryotypes for chromosome arrangements (inversions or translocations) have reduced fertility due to the production of chromosomally unbalanced gametes.

In terms of the selection cases in Table 3-2, how would you classify selection on these gene arrangements?

3.2.2 Common Genetic Diseases and Malaria

The explanation for the maintenance of the high *HbS* frequency in populations in Africa and Asia resides in its interaction with the malarial parasite *Plasmodium falciparium* (Allison, 1954). The evidence, reviewed extensively by Vogel and Motulsky (1979), suggests that *HbA*/*HbS* heterozygotes are far less likely to die of malaria than are the normal homozygotes. The reason for this seems to be the occurrence of the sickle cell trait in *HbA*/*HbS* individuals. In these individuals, when there is oxygen stress in the capillary systems of the body the hemoglobin in the red blood cells crystallizes and the cells take a sickle shape. This may cause small ruptures of the cell membrane, which in turn causes a leaking of the potassium-rich cytosol. Since the parasite lives inside the red blood cells and requires a high potassium concentration to thrive, the link between sickling and parasite control is established (Pasval et al., 1978). Together with the very strong selection against the anemic homozygote *HbS*/*HbS*, the amelioration of the disease in sickle cell (*HbS*/*HbA*) individuals produces heterozygote advantage in childhood viability of the kind in case 4 of the previous section. A number of population studies suggest that this selection is strong enough to explain the polymorphism.

There is no direct evidence for such selection as the basis of *HbC* and *HbE* polymorphism, although both occur in heavily malarial areas. For *HbE*, the suspected mosquito is *Anopheles minimus*, whose distribution in Southeast Asia is coincident with that of *HbE*. Similarly, heterozygotes for *HbT* (homozygotes which have B-thalassemia, section 1.2.2) appear to afford protection against malaria (Motulsky, 1975), and *G6PD*-deficient cells produce a lower concentration of reduced glutathione in their red cells, thereby inhibiting growth of the parasites (Luzatto et al., 1969).

These three examples invoke an advantage to heterozygotes in that one homozygote is subvital because of the genetic disease and the other homozygote suffers a higher death rate because of a disease present in the environment, a disease that is less severe in the heterozygote. The overdominant selection is dependent on malaria being a cause of excessive death for the "normal" homozygote. If the environmental disease is eradicated or the population is moved to another environment without the disease, then the difference between the "normal" homozygote and the heterozygote may vanish or perhaps reverse, with the result that the population is left with a disease allele in high frequency. Subsequently, directional selection would work to eliminate the disease allele and we would expect it to disappear given enough time.

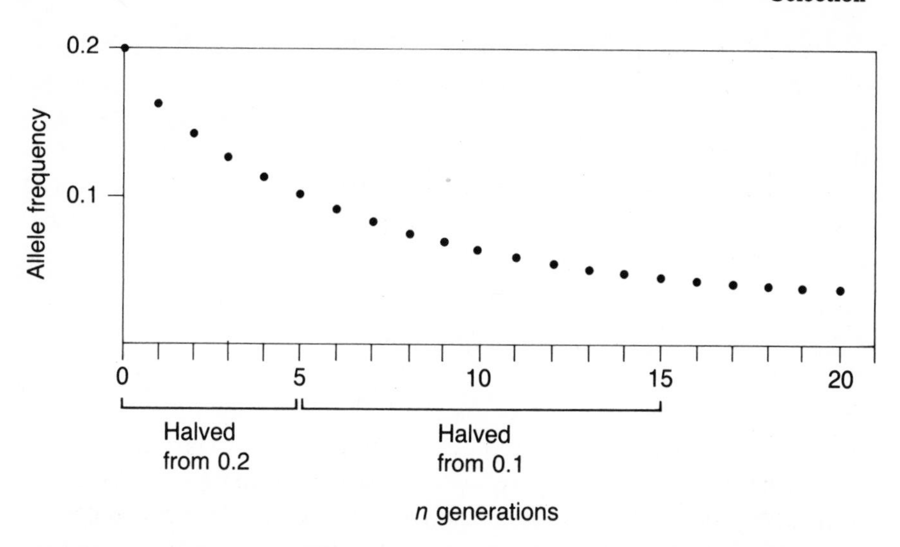

Figure 3-8. Decrease in frequency of a recessive lethal allele starting at a frequency of 0.2. The number (n) *of generations it takes for the population to reach half of the original allele frequency increases as* n = *1/*q *with decreasing frequency* (q) *of the recessive.*

Let A be the normal allele and a the recessive disease allele and suppose that the genotypic fitnesses are 1, 1, $1 - s$ of *AA, Aa, and aa* after the enviromental disease has disappeared. Let p and q be the initial allele frequencies of A and a (the equilibrium allele frequencies with the disease present), and to simplify the argument consider the most extreme case of a lethal recessive disease, that is, $s = 1$:

$$q' = \frac{q}{1 + q}.$$

After n generations, the frequency of the disease allele is

$$q^{(n)} = \frac{q}{1 + nq},$$

so that the allele frequency of a converges towards 0. This convergence is very slow, however, because the presence of the bad allele in heterozygotes, who are just as fit as the normal homozygote, delays its loss (Figure 3-8). Suppose $q = 0.2$; then the allele frequency reaches 0.1 after five generations, 0.05 (the *Hbs* frequency in U.S. blacks) after an additional 10 generations, and 0.025 after yet another 20 generations. The convergence becomes slower and slower and the genetic disease remains in the population at rather high frequencies for a long time. The average incidence of the disease in the first 35 generations of the preceding example is about 4 per 1000 births, which is much higher than the usual total incidence of all other single-gene diseases in a human population (Table 1-7).

3.2.3 Maternal-fetal Incompatibilities

As mentioned in section 1.2.2, some blood group polymorphisms can produce disease-causing immunological interactions for certain mother-child combinations. These reactions place heterozygous offspring of homozygous mothers at risk, resulting in a decreased survival probability of a heterozygous fetus compared to a homozygous fetus, everything else being equal. For a two-allele polymorphism such as *D* and *d* of the Rhesus system, this results in effectively underdominant selection where only stable monomorphic allele frequency equilibria are possible. Thus the existence of the Rhesus polymorphism presents us with an evolutionary paradox.

The paradox remains even if we take into account that the zygotic viability of the heterozygote *Dd* depends on the frequency with which the mother-child combination *dd-Dd* occurs. If the average viability of a *Dd* fetus in a *dd* mother is $v = 1 - s$ relative to the average viability of fetuses in all other mother-child combinations, then from Table 2-5 the genotypic fitnesses are $v_{DD} = 1$, $v_{Dd} = 1 - spq^2$ and $v_{dd} = 1$, so $q' > q$ when $q > 1/2$ and $q' < q$ when $q < 1/2$, but the convergence of the frequency of *D* to 0 for $q > 1/2$ is slow and the convergence of q to 0 for $q < 1/2$ is even slower, since in both cases v_{Dd} becomes close to v_{DD} or v_{dd} as one of the alleles becomes rare. However, only in Mongoloid populations, where $v_{Dd} = 1 - s \times 0.002$, can the selection in nature against allele *d* be said to be small (Table 1-8); in Caucasoid ($v_{Dd} = 1 - s \times 0.09$) and African ($v_{Dd} = 1 - s \times 0.02$) populations, the selection is not entirely negligible because s may have been rather large until recently. (For more discussion on this paradox, see Feldman et al., 1969.)

3.2.4 More Complex Selection on One Gene

The simple model of natural selection analyzed in section 3.2.1 can be extended in many ways. The *Rh* system of section 3.2.3 is an example in which the viabilities were allowed to depend on the genotypic composition of the population. Further extensions include other components of natural selection (see section 2.1.1). In general, the results of section 3.2.1 hold for any two-allele polymorphism if zygote to adult viability may be replaced by a constant individual *fitness* value, which is a measure of the individual's performance during its whole lifetime. Roughly speaking, the fitness of an individual zygote should be its expected number of offspring zygotes in the next generation. For human populations, two complications arise in the calculation of fitness. First, human populations cannot be separated into discrete generations, as the individuals may breed from the time maturity is reached at more or less irregular intervals and for an extended time. Two individuals that leave the same number of offspring can very well differ in fitness if one breeds early and the other late. For example, if both produce a single offspring, one at age 15, the other at age 30, the offspring of the early breeder could conceivably breed at the same time as the late breeder, giving the early breeder effectively more than one offspring. A demographic description of the various genotypes is necessary to overcome this problem in the definition of fitnesses (see Cavalli-Sforza and Bodmer, 1971). The second and main problem in the definition of a fitness value is that certain aspects of natural selection cannot be ascribed to the individual; fecundity, for example, describes a fitness attribute of the mated male and female pairs, while sexual selection will usually depend on the genotypic composition of the population. These interactive contributions to fitness can produce evolutionary results which differ from those in section 3.2.1 (Feldman et al., 1983), but by and large the principles outlined there can be expected to hold in some generality.

Exercise 3.2.D

The HLA antigen *A1* is strongly associated with psoriasis in the Japanese population (Table 1-10). Suppose (only for this exercise) that the probability that psoriasis patients become parents is 90 percent of that for others, and suppose that the fecundity of parents is not affected by the disease.

Now reduce the HLA-*A* locus variation to two alleles: the allele A_1 produces antigen $A1$ and allele A_2 is responsible for all non-$A1$ antigens. (A_2 includes all alleles other than A_1 at the A locus).

a. Calculate the fitness value of genotype A_1A_1 relative to A_2A_2 corresponding to the selection induced by the psoriasis association in the Japanese population.
b. Would induced selection of this kind have an important effect on the frequency of antigen $A1$ in the Japanese population? (See Table 1-9).

Exercise 3.2.E

Many species of sulfur butterflies show a sex-limited genetic polymorphism. The females show two phenotypes: a white *alba* and a normal coloring (yellow, orange or red), which we may call *typica*. The males show the *typica* phenotype. The phenotypes are determined by two alleles, A dominant and a recessive, at an autosomal locus, and *alba* is the dominant phenotype expressed only in females.

In an experiment with *Colias alexandra,* pairs of *alba* and *typica* females were placed in an area where males were flying, and the number of male visits to the two female phenotypes was observed to be

Alba	*Typica*	Total
25	118	143

From Graham et al. (1980).

a. With this information, how would you predict the changes in phenotypic frequencies in the population?

In the studied population in the Rocky Mountains, the frequencies among the females of *alba* and *typica* were 0.126 and 0.874, respectively.

b. Can these phenotypic frequencies be used to estimate the allele frequencies? Why not?

The *alba* females were shown to be in better physical condition than the *typica* females, suggesting that they would survive longer and produce more offspring.

c. Does this information help to explain the polymorphism?

Deviations from random mating have special effects on loci where the genetic variation is subject to selection. Even under purely zygotic selection, the simple correspondence between the ordering of viabilities and the evolutionary dynamics in Table 3-2 breaks down. Deviations from random mating may create sexual selection, if the deviation results in different probabilities of mating for the various genotypes. For instance, assortative mating, as described in section 2.4.2, may result in sexual selection if the fraction of assorting females differs among the genotypes (unless the mating is strictly monogamous). Consider a polygamous species in which a certain fraction of the females of a given genotype chooses mates of the same genotype, and the rest of the females of that genotype choose a mate at random from among all males in the entire population. This mating system will favor males of the genotype that has the highest rate of assortment, because these males have the highest chance of mating with an assorting female yet have the same chance as any other male to mate with a randomly mating female (Feldman and Christiansen, 1984). Many other models of assortative mating have been studied, and virtually all result in some form of sexual selection (Karlin, 1968; O'Donald, 1980).

Deviations from random mating due to population subdivision can create a balance between the convergence to specific equilibria in local populations and the smoothing effect of migration, which may work in another direction. As an example, consider Wright's island model (section 2.4.3) and assume underdominant selection (selection case 3, Table 3-2). Here an isolated population may converge to either of the stable monomorphic equilibria, and if, for historical reasons, the subpopulations were initially at different allele frequencies, they may become monomorphic for different alleles. A small amount of migration does not change this situation much; each population will be close to a monomorphic equilibrium, balanced between migration and selection. If, on the other hand, migration is frequent, the population will become homogeneous and monomorphic for the same allele (Karlin and McGregor, 1972). Thus even if selection is homogeneous throughout the population, considerable variation in allele frequencies may result if migration is small compared to the fitness differences. If selection (of any type) varies among the local populations, an analogous difference in evolutionary outcome between high and low migration occurs. For low migration (compared to the fitness differences among genotypes), each local population will be close to a stable equilibrium determined by the local selection. For high migration, the allele frequencies in the population are mainly determined by some kind of average selection over the population and little geographic differentiation in allele frequencies is expected.

3.2.5 Selection with More Loci

The fitness of a given genotype with respect to the allelic variation at a locus may depend on the genotypes with respect to the variation at other loci. For example, the Rhesus mother-offspring incompatibility reaction occurs less often in mothers with blood type *O* and most often for mothers with blood type *AB*. Thus, the increased mortality of *Dd* children from *dd* mothers induces fitness differences between the genotypes at the locus for the *ABO* blood groups.

A theory similar to that in section 3.2.1 has been developed for simultaneous variation at two loci. This theory, however, is far more complicated mainly because, as described in section 2.3, convergence to Robbins proportions is not instantaneous and depends on the recombination frequency. Except for very special types of selection, the viability differences between two-locus genotypes will create linkage disequilibrium in every generation. Recombination causes this linkage disequilibrium to decrease every generation but not necessarily to zero. We can expect a balance between the increase by selection and the reduction by recombination in the deviation of the gamete frequencies from Robbins proportions. For any given set of genotypic viabilities that can maintain a polymorphism at both loci, the equilibrium value of the linkage disequilibrium will depend on the recombination frequency c. For c close to 1/2, strong selection is required to maintain any appreciable linkage disequilibrium. When c is small, weaker selection may preserve the association between the loci. The maintenance of polymorphism at both loci for given fitness values may even depend on c. For a recent review of theoretical results for two-locus selection, see Ewens (1979).

Selection on the allelic variation of a gene may influence the variation at linked yet selectively neutral genes through a phenomenon called *hitchhiking*. As an illustration, consider again the *HbA-HbS* variation. Suppose the human population were initially monomorphic *HbA/HbA*, but at a certain time a mutant *HbS* became established in a population exposed to endemic malaria. This mutant must have occurred in one copy of the gene coding for the hemoglobin β-chain, so that initially only one gamete in the whole population carried the *HbS* allele. Therefore the linkage disequilibria with alleles at all other loci were maximal. Let $1 - s$, 1, $1 - t$ be the fitnesses of *HbA/HbA*, *HbA/HbS*, and *HbS/HbS*. Then the change in the allele frequency of *HbS* may be approximated initially by

$$q' = \frac{q}{1 - s}.$$

Now consider a linked gene with selectively neutral alleles B_1 and B_2 in frequencies p_1 and p_2, and let x be the frequency of the gamete $B_1\,HbS$. Suppose $D = x - p_1 q$ is initially different from 0. The gamete frequency x changes approximately as according to

$$x' = \frac{(1 - c)x + cp_1 q}{1 - s},$$

and we may assume that the initial linkage disequilibrium $D = x - p_1 q$ is positive. If the change in the frequency of B_1 is initially small, then

$$D' \approx x' - p_1 q' = \frac{D(1 - c)}{1 - s}$$

and the linkage disequilibrium between the two loci will initially increase if $c < s$. Thus loci sufficiently closely linked to the hemoglobin β-locus will reflect an initial association of an allele to the increasing *HbS* allele. An allele initially associated with *HbS* will increase along with *HbS* and at the same time maintain the association.

This phenomenon may be the explanation for the observed association between the *HbS* allele and a restriction enzyme recognition site in African populations (section 1.2.6). The phenomenon may even account for some of the isolated high occurrences of recessive genetic diseases described in section 1.2.2. The hitchhiking effect of an increase in a favored allele can be strong enough to carry along recessive lethals to rather high frequencies. Selection on rare recessive lethal alleles is very slight in this respect, as a recessive disease with a frequency of 0.0001 has a gene frequency of 0.01 from the Hardy-Weinberg proportions, so the average fitness of the lethal allele is 0.99 relative to 1 for the normal allele.

A rapid change from a very low to a rather high frequency of a selectively favored allele, as in *Biston betularia* (Figure 3-7), may well result in hitchhiking by recessive deleterious alleles. The favored dominant allele *A* would then be associated with deleterious alleles more often than the recessive allele *a*. Genotype *AA* then has a viability that is lower than *Aa* because it is more often homozygous for a recessive deleterious allele. The heterozygote *Aa* has a higher viability than *aa*, because it is of the favored phenotype. Thus, as long as the association between *A* and the hitchhiking deleterious alleles persists, the viabilities of the genotypes *AA*, *Aa*, and *aa* are overdominant (Table 3-2, case 4), and the population will converge to a polymorphic equilibrium and not to monomorphism for *AA*. This phenomenon is called *associative overdominance* (Frydenberg, 1964).

4

Phenotypes and Genes

The genotypic contributions to all heritable phenotypic variation must obey the basic genetic rules of Chapter 2. For many important phenotypes, however, the role of genetics is not known and it is not clear how to incorporate these basic rules. The polygenic model of section 1.1.3.2 is useful in this context because it forms the basis for *simulation models* of the inheritance of phenotypic variation, using the simple population genetic rules. The predictions of these simulation models allow us to ask, for example, whether the polygenic model is sufficient to describe an observed pattern of population variation and familial aggregation. In this section we indicate how this sort of question may be addressed, especially in the context of human genetics. Most of the technical details are beyond the scope of this book and are therefore omitted.

In Chapter 1 we summarized the array of possible phenotype distributions in populations and the way in which these may aggregate in families. If the trait segregates within families, according to Mendel's rules, then the usual procedure is to accept the trait as genetic even if no specific gene product has been identified. Many attempts have been made to clarify the role of genetics in determining traits that do not exhibit Mendelian segregation but show some familial aggregation. Familial aggregation may occur either as concordance among monozygous twins, as increased numbers of affected people among relatives of the proband compared to unrelated controls, or simply as correlation among relatives (section 1.3.3). The method of evaluating the role of genetics usually involves the construction and analysis of statistical models that represent the phenotype as the outcome of the joint action of genes and environment. If it can be inferred that genes are important for the trait under the conditions of the model, then perhaps the likelihood of finding specific gene products that affect the disease is increased. On the other hand, even if such biochemical markers are found, intervention (with the exception of genetic counseling) is environmental, as exemplified by the dietary control of phenylketonuria symptoms.

Without going too deeply into statistical methodology, we will proceed to outline the main ideas used in modeling diseases with complex inheritance patterns.

4.1 Genetic Models of Quantitative Variation

The statistical model underlying much of the inference and prediction made about genetic causes of phenotypic variation was devised by R. A. Fisher (1918). The model allows a genetic interpretation of observed variation and familial aggregation of a trait in a population under certain simplifying assumptions. It also suggests how models of phenotypic variation due to a single gene may be generalized to depict phenotypic variation with polygenic inheritance.

4.1.1 The Standard Model for Quantitative Variation

Consider a quantitative character that exhibits variation in a population which can be described by a normal distibution with mean m and variance V_P, the phenotypic variance. Suppose that a gene with alleles A_1 and A_2 influences this trait. Let Γ_{11}, Γ_{12}, and Γ_{22} be the *genotypic values*, i.e., mean values of the trait among individuals of the genotypes A_1A_1, A_1A_2, and A_2A_2 (Figure 4-1), and let $\gamma_{11} = \Gamma_{11} - m$, $\gamma_{12} = \Gamma_{12} - m$ and $\gamma_{22} = \Gamma_{22} - m$ be the *genotypic effects* of the genotypes at locus A. If there is random mating so that the Hardy-Weinberg proportions are valid, then

$$m = p_1^2\Gamma_{11} + 2p_1p_2\Gamma_{12} + p_2^2\Gamma_{22},$$

or, equivalently,

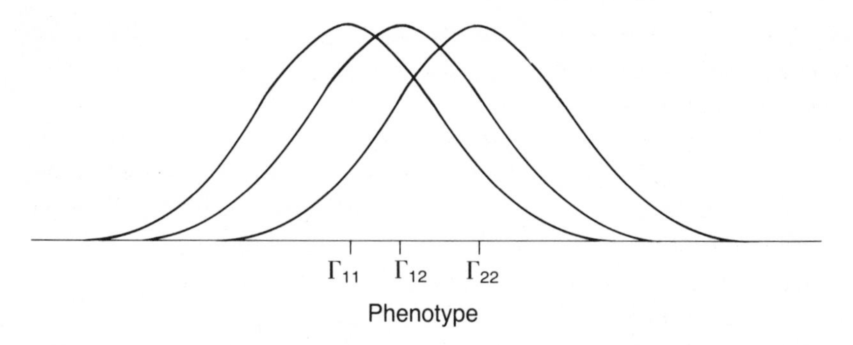

Figure 4-1. The genotypic values and the phenotypic distribution for given genotypes corresponding to the genotypes A_1A_1, A_1A_2 *and* A_2A_2 *(compare with Figure 1-29, page 54).*

$$p_1^2\gamma_{11} + 2p_1p_2\gamma_{12} + p_2^2\gamma_{22} = 0.$$

The variance among the genotypes, the *genotypic variance*, is

$$V_G(A) = p_1^2\gamma_{11}^2 + 2p_1p_2\gamma_{12}^2 + p_2^2\gamma_{22}^2.$$

Now assume that the character in question shows polygenic inheritance, and suppose it is independently influenced by many gene loci, A, B, C, etc., so that the genotypic effect of a multilocus genotype is the sum of the genotypic effects at each locus. Assume further that the total genotypic variance is the sum of the genotypic variances at each locus:

$$V_G = V_G(A) + V_G(B) + V_G(C) + \cdots .$$

This assumption is fulfilled for additive loci in the preceding sense if the multilocus genotypic proportions are given by Robbins proportions.

The total variance V_P in the population is the sum of the genotypic variance V_G and a residual variance V_E called the *environmental variance*:

$$V_P = V_G + V_E.$$

The environmental variance is the variance in trait value between individuals of the same genotype.

Typically, only m and V_P are observable in the population. In particular, we do not know the loci involved in the genotypic contribution to the character. We can, however, observe familial aggregations and the correlations among relatives. To interpret these observations, we need to consider theoretically the observable effects of a single model locus and then to use the additive model to generalize these single-locus results into observable effects.

The *average effect* α_1 of the allele A_1, say, is the average deviation from the population mean of the genotypes in which this allele is found, that is, for the two-allele case

$$\alpha_1 = p_1\gamma_{11} + p_2\gamma_{12},$$

$$\alpha_2 = p_1\gamma_{12} + p_2\gamma_{22}.$$

According to this definition, the average effect of all alleles at a locus is zero, that is,

$$p_1\alpha_1 + p_2\alpha_2 = 0.$$

If the contribution of the gene with alleles A_1 and A_2 were due entirely to additive contributions of the alleles, then we would expect the genotypic effects to be $2\alpha_1$, $\alpha_1 + \alpha_2$, and $2\alpha_2$ for A_1A_1, A_1A_2, and A_2A_2. The variance due entirely to the allelic effects is then

$$\begin{aligned} V_A(A) &= 4p_1^2\alpha_1^2 + 2p_1p_2(\alpha_1 + \alpha_2)^2 + 4p_2^2\alpha_2^2 \\ &= 2(p_1\alpha_1^2 + p_2\alpha_2^2). \end{aligned}$$

This variance is called the *additive genetic variance* and the residual genotypic variance, V_D, given by

$$V_G(A) = V_A(A) + V_D(A),$$

is called the *dominance variance*. It is expressed as the variance of the residual genotypic effects, $\gamma_{11} - 2\alpha_1$, $\gamma_{12} - \alpha_1 - \alpha_2$ and $\gamma_{22} - 2\alpha_2$, obtained when the allelic effects have been taken away. With this partitioning of the genetic variance at each contributing locus, we have

$$V_P = V_G + V_E = V_A + V_D + V_E.$$

This partitioning of the phenotypic variance was made in terms of a two-allele model. However, all of the foregoing expressions pertaining to variation at one locus may readily be generalized to include alleles $A_1, A_2,..., A_k$ with frequencies $p_1, p_2,..., p_k$ and with the genotypic frequencies in Hardy-Weinberg proportions.

Exercise 4.1.A

Consider a population at Hardy-Weinberg equilibrium with respect to the variation at an autosomal locus with the alleles A_1 and A_2 at the frequencies 0.25 and 0.75. Let the genotypic values for some character be

A_1A_1	A_1A_2	A_2A_2
16	48	80

a. Find the population mean, the genotypic effects, and the genotypic variance.

b. Find the average effects of A_1 and A_2, and compute the additive variance.

c. What is the value of the dominance variance?

These theoretical calculations may be illustrated by using the distribution of red cell acid phosphatase activities within electrophoretic phenotypes (Figure 1-29, page 54) to partition the phenotypic variance of the population in Figure 1-4 (page 5) into components. After the activities are transformed by taking their logarithms, the genotypic variance due to the genotypes at the electrophoretically defined locus is 66 percent of the phenotypic variance. The remaining 34 percent of the variance is due either to environmental variance or to variation at other gene loci. The genotypic variance due to the acid phosphatase locus may be broken down further into an additive and a dominance component. The genotypic variance in this example, however, is due almost entirely to additive allele effects, so that V_A/V_P, V_D/V_P and V_E/V_P take the values 0.65, 0.01, and 0.34, respectively, in the log-transformed data. Clearly a large percentage of the population variation of Figure 1-4 is explained by the genotypic variance at the electrophoretic locus. This example illustrates the fact that a unimodal quantitative character can be strongly influenced by a single locus, but the example also shows that there are no immediate features of the phenotypic distribution that would lead us to that conclusion. Only genotypic information such as that presented in Figure 1-13 allows the contribution of a given locus to the variation in a character to be evaluated.

It is worth emphasizing that the standard model produces a *description* of the variation of traits in a given population at a given time. The model partitions the phenotypic variance into components according to the prevailing genetic and environmental conditions. The strength of the model is that the variance components can be estimated from the pattern of familial aggregation under certain simplifying assumptions. The model, however, only describes the *variation* in the population. The absolute level of the trait value as described by the population mean is neglected in this description. The value of the mean can be changed by changing the environment as, for example, the increase in mean height of humans by improved nutrition or the increase in mean IQ under adoption, or it can be changed by changing the genetic constitution of the population as, for example, the increase in yield from livestock obtained by selection. These changes in the mean may occur without appreciably changing the total phenotypic variance or its components.

4.1.2 Genetic Correlations Between Relatives and Heritability

The foregoing model allows the evaluation of resemblance between relatives using techniques similar to those outlined in Chapter 2 under similar assumptions of random mating and no selection. That is, we assume Hardy-Weinberg equilibrium at each of our model loci and linkage equilibrium among all loci. With these assumptions, we expect the phenotypic distribution to be the same each generation (Figure 1-14) if the population environment is constant.

The familial resemblance predicted from the standard model of quantitative inheritance is usually given in terms of expected correlations between relatives (the value r in section 1.1.3.1) or as the corresponding covariances (simply the value rV_P). A list of theoretical covariances between the most commonly available relatives is presented in Table 4-1. From the algebraic relationships between these covariances or the resulting correlations, it is sometimes possible to estimate the relative contributions of the different variance components to the phenotypic variance. The ratio $h^2 = V_A/V_P$ is called the *heritability* of the trait (termed *narrow sense heritability* because it does not include all of the genetic contributions to the phenotypic variance). This heritability may, for example, be estimated from the covariance between offspring and one parent as illustrated by the mother-offspring relation in Figure 1-15. The data in this figure may be described by a mother-offspring regression, and the regression coefficient, the slope of the line, is 0.4. This slope is calculated as the ratio of the covariance between the maternal and corresponding offspring values to the variance of the mother values. Thus from Table 4-1 the slope

Table 4-1. *Theoretical Covariance in Phenotype Between Relatives.*

Degree of relationship	Covariance
Offspring and one parent	$V_A/2$
Offspring and average of parents (midparent)	$V_A/2$
Half siblings	$V_A/4$
Full siblings	$(V_A/2) + (V_D/4)$
Monozygotic twins	$V_G = V_A + V_D$
Nephew and uncle	$V_A/4$
First cousins	$V_A/8$
Double first cousins	$(V_A/4) + (V_D/16)$

is expected to be $(V_A/2)/V_P$, which produces the heritability as twice the slope, $h^2 = 0.8$. The data of Table 1-1 (page 19) provide a more detailed description of the genetic variance because if the sib-sib and parent-child correlations are both available, then both V_A and V_D may be estimated (Table 4-1). If we assume that the phenotypic variance in parents equals that of offspring, then the parent-child correlation is $(V_A/2)/V_P$ while the sib-sib correlation is $(V_A/2 + V_D/4)/V_P$. Thus, these two correlations taken from Table 1-1 provide estimates of heritabilities of about 0.4, 0.5, and 0.6 for diastolic blood pressure, height, and weight, respectively. The corresponding estimates of V_D/V_P are about 0.4, 0.1 and 0, respectively. This estimation of h^2 and V_D/V_P from the data of Table 1-1 is, of course, very imprecise. The rough spirit of this analysis is reflected in the naive comparison between the theoretical and observed correlations from Table 4-1 and Table 1-1. A full statistical analysis of the available family data would provide more reliable estimates.

Exercise 4.1.B

a. Use Table 2-5 to calculate the covariance in genotypic values between mother and offspring using the formula

$$\sum_{i=1}^{3} \sum_{j=1}^{3} \gamma_i \gamma_j \, P\,(\text{mother of genotype } i \text{ and offspring of genotype } j)$$

where i or j = 1, 2, and 3 corresponding to the genotypes A_1A_1, A_1A_2, and A_2A_2, respectively. (Convince yourself that this is in agreement with the correlation definition in Section 1.1.3.1.)

b. Show that this covariance is equal to $\frac{1}{2} V_A$.

Exercise 4.1.C

a. Use the result of Exercise 2.2.C to calculate the covariance in genotypic values between sibs as

$$\sum_{i=1}^{3}\sum_{j=1}^{3} \gamma_i\gamma_j\, P(\text{sib pair with genotypes } i \text{ and } j),$$

where i or $j = 1, 2,$ and 3 corresponding to the genotypes A_1A_1, A_1A_2, and A_2A_2, respectively. (Convince yourself that this is in agreement with the correlation definition in section 1.1.3.1.)

b. Show that this covariance is equal to $\frac{1}{2}V_A + \frac{1}{4}V_D$.
(*Hint:* write γ_i in terms of additive allele effects, e.g., $\gamma_1 = 2\alpha_1 + (\gamma_1 - 2\alpha_1)$, and use this to simplify the expression for the covariance.

Heritability is economically important in animal breeding because under some controlled experimental conditions available to animal breeders, it predicts the response to artificial selection (see Falconer, 1981). Figure 4-2 shows the results of selection for high and low growth rate in an experimental population of mice. The slope of the response curve for the change in growth rate from generations 1 through 6 is expected to be proportional to the heritability of the trait, with the proportionality constant given by the selection pressure used. The outcome of such a selection experiment may therefore be predicted via estimates of the heritability from data on familial aggregation of the trait. Data from such selection experiments also provide an estimate of the heritability, known as the *realized heritability* of the trait. The realized heritability may differ from the familial heritability estimate for a variety of causes. For example, the applied selection pressure, the artificial selection, may be enhanced or opposed by natural selection on the trait. The familial heritability may be influenced by common environmental effects, such as maternal effects on related individuals. Finally, random genetic drift in the usually small experimental population may decrease the effectiveness of artificial selection and thereby decrease the realized heritability by retarding the response to selection.

The correlations among relatives, and the heritability, refer to the transmission of the trait within the population in which they are measured. Thus a phenotypic difference between two populations with respect to a certain character is not well described by the familial correlations within the populations, that is, *the population difference is not in any sense genetically comparable to the difference among individuals within a population.*

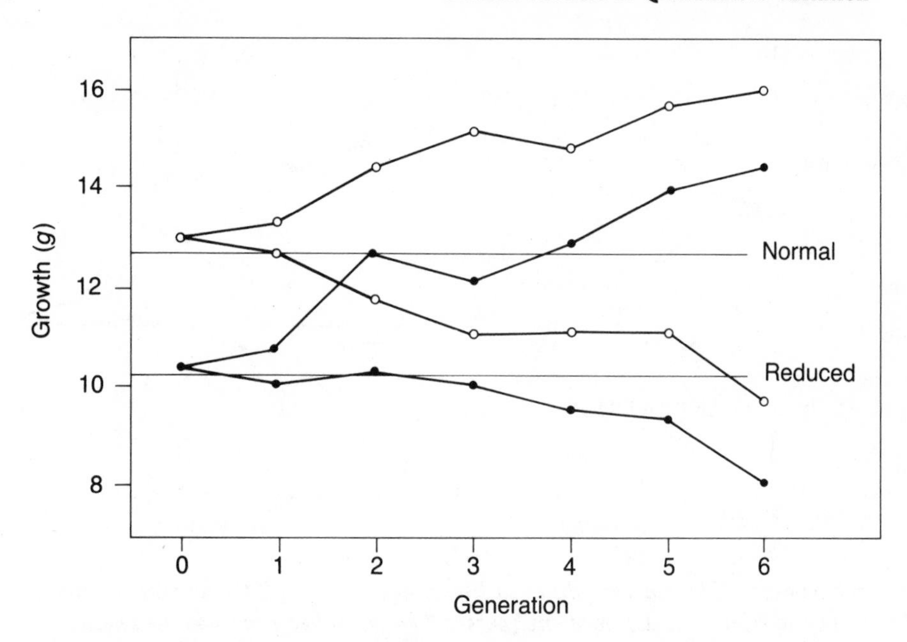

Figure 4-2. Graphs showing the response of a mouse population to selection for higher and lower growth (the character in Figure 1-8). The selection experiment was performed in two environments: one in which the mice were given food with a normal *protein content and one in which the feed had a* reduced *protein content. The lines marked with normal and reduced show the mean growth rate in an unselected population in the two environments. The mean growth in populations selected for high and low growth are shown as functions of the number of generations that selection was applied (○, normal diet; ●, reduced diet). (From unpublished data of V. Nielsen.)*

4.1.3 Other Sources of Familial Correlations

For the quantitative traits described in section 1.1.1, there are additional reasons why heritability is of dubious value as a measure of the importance of genes in determining the variation among human phenotypes. The following list is not in any specific order; for most traits the order of significance is not known.

1. *Correlated environments.* The comparison of the correlations among relatives with expectations from the model ignores any correlation among environmental contributions to the trait values in the relatives. Thus we might consider diet to be an important contributor to blood pressure. Of course, diets of parents and offspring or sibs are expected to be very similar. Such similarities are ignored in the foregoing treat-

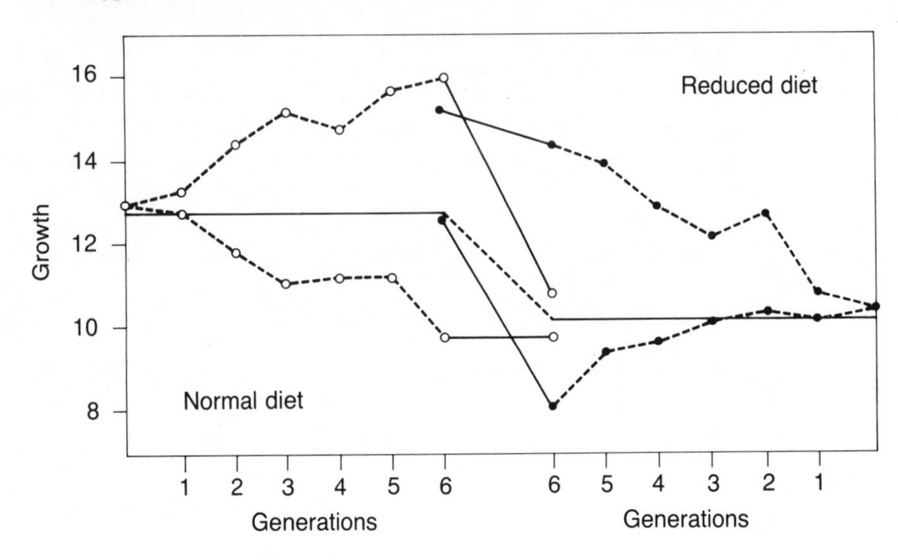

Figure 4-3. The same graphs as in Figure 4-2, with (left) the normal environment and (right) the deprived environment. In the center growth in the selected populations in generation 6 is shown in the environment in which they are selected compared with their growth in the other environment. When grown in the reduced environment, the populations selected on the normal diet show only a very small deviation from the growth in the unselected population. Similarly, the low selection population from the reduced environment is close to the unselected population when grown in the normal environment. Only the high selection population from the reduced environment shows any appreciable response when grown in the normal environment. Selection for higher or lower growth rate in the two environments gives comparable phenotypic responses, but the genetic change in the population producing these responses must be different in the two environments. Genotypes that produce higher growth rate on a normal diet do not necessarily produce a higher growth rate on a reduced diet. (From unpublished data of V. Nielsen.)

ment. In particular, it is difficult to separate the increase in sib correlation due to the contribution of such correlated environments among sibs from the increase due to dominance terms.

2. *Genotype-environment interaction.* In many organisms there are known interactions between genotypes and environments (Figure 4-3). The biochemical functions of the products of different genotypes may depend on the environment in which they occur. Figure 4-4 depicts such a case of genotype-environment interaction in terms of the norm of reaction of genotypes. Thus the heritability is only descriptive in

the environment in which the various measures are taken. It does not relate to a possible change in the phenotype mediated by a change to another environment, as might occur in many treatment situations. This consideration is particularly relevant to studies of the familial aggregation of behaviors, personality traits, and IQ. For some such traits, differences among populations have been observed, but in contrast to genetic differences among groups (for example, skin color or blood group gene frequencies), which may be demonstrated unambiguously, the between-group differences in a quantitative character do not bear upon the issue of the genetic determination of the phenotype under study, although the contrary has often been claimed.

3. *Assortative mating*. Assortative mating is the term used to describe nonrandom mating due to positive correlation between the phenotypes of mates. This correlation could be the result of mate selection based on the phenotype under study, as might be the case if the familial aggregation of height were the topic of interest. Alternatively, it might be the result of environmental stratification. For example, if IQ is the trait under study, then assorting based on educational achievement could be important. In fact, it turns out that the correlation between mates is high for educational level, and this must be taken into account in the variance breakdown. Assortative mating increases the genetic variance over that expected with random mating under a variety of assumptions. But if the mating is "selective," so that the variance in phenotypes among mating couples is less than that in the overall population, then the overall variance is decreased (Karlin, 1979). It appears that the different ways of specifying the mode of assortment lead to different consequences for the expected familial aggregation (see, for example, Feldman and Cavalli-Sforza, 1979a).

These admonitions concerning the genetic interpretation of familial aggregation reflect two basic properties of our analysis of quantitative traits: (1) the observations on transmission of a trait are purely descriptive and refer specifically to the circumstances in which the observations are made, and (2) the genetic model underlying the interpretation is extremely simplistic and neglects many population genetic phenomena known to be of relevance in the understanding of simple genetic polymorphisms (see, for example, Feldman and Lewontin, 1975). With this in mind, it is truly impressive how often the quantitative genetic approach can be helpful, but it is not surprising that such an approach may turn out to be misleading if it is not used with care.

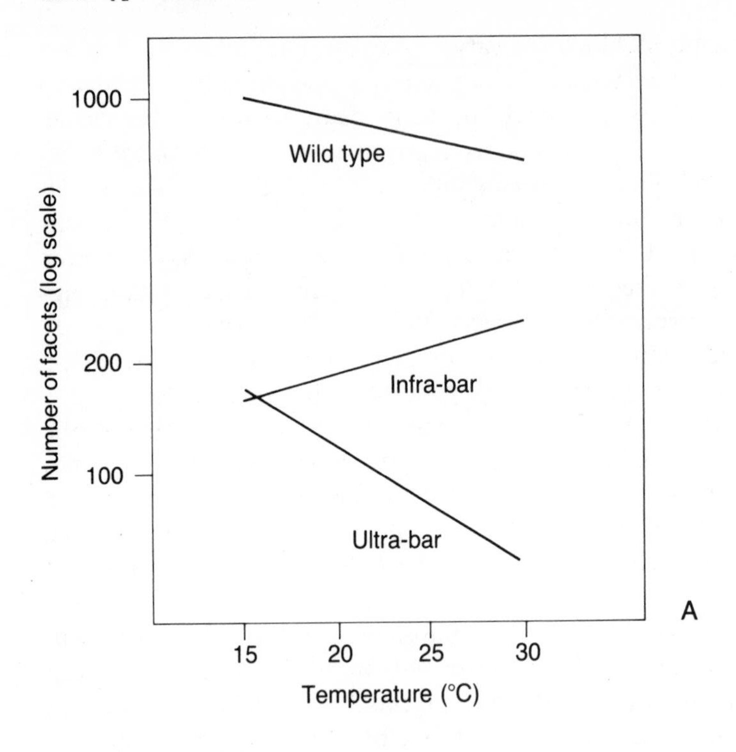
1000
200
100
Number of facets (log scale)
Wild type
Infra-bar
Ultra-bar
A
15
20
25
30
Temperature (°C)

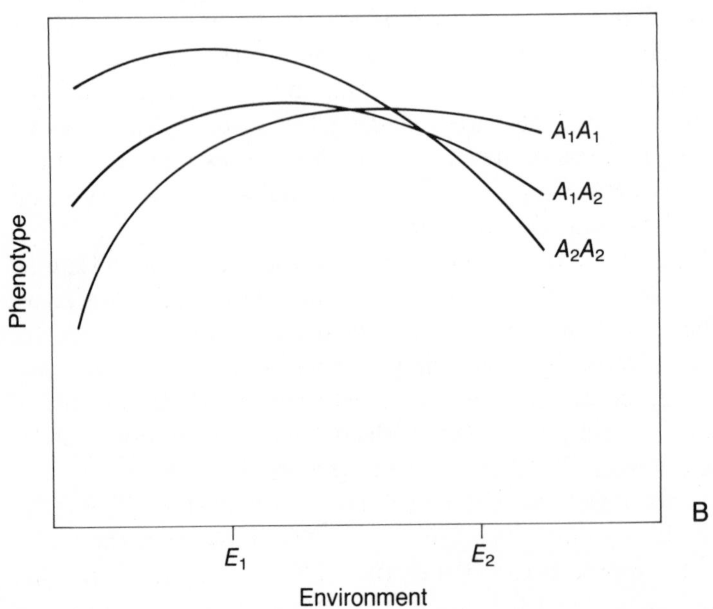
Phenotype
A_1A_1
A_1A_2
A_2A_2
B
E_1
E_2
Environment

4.1.4 More General Linear Transmission Models

The model of section 4.1.1 may be formally expressed in terms of a linear relationship between the phenotypic value P, the genotypic value Γ, and the environmental deviation E of the individual:

$$P = \Gamma + E.$$

The model of genetic transmission is then specified in terms of the relationship between Γ_O, the genotypic value of an offspring, and the genotypic values of the parents Γ_M and Γ_F, where the subscript M refers to the mother and F to the father. Suppose now that we are considering an additive character where $V_G = V_A$ that is, the genotypic effects at each of the relevant loci are given by $\gamma_{ij} = \alpha_i + \alpha_j$ and $V_D = 0$. In this case, a simple formulation of the transmission model is

$$\Gamma_O = \frac{1}{2}(\Gamma_M + \Gamma_F) + x,$$

where x is a random variable describing the segregation within a group of sibs: The segregation deviation x of the individual offspring is

Figure 4-4 (opposite). A: *Mean number of eye facets in* Drosophila melanogaster *in wild type flies and flies homozygous for two different alleles at the* Bar *locus as a function of the temperature during their development. All three genotypes change their genotypic value with temperature. Curves describe* norm of reaction *to temperature of the three genotypes. Because of their different norms of reaction, the two mutant genotypes, which are phenotypically distinct at high temperatures, are indistinguishable at low temperatures. In general, different genotypes can show different norms of reaction to a given environmental variable.* B: *A theoretical example shows the possibility of reversal in the ordering of the genotypic values when going from environment E_1 to E_2 at the same time as the genotypic variance decreases. The environment may vary among the individuals in a population, and the norm of reaction model may be used to judge the effect of this variation. Variation among individuals in temperature will clearly produce a larger variance in the mutant homozygotes than in the wild type shown in* A. (A *after Suzuki, Griffiths, and Lewontin, 1981).*

assumed to be normally distributed with mean 0 and variance $V_G/2$. Our simple model describing the transmission on the phenotypic level is then fully specified by assuming that the environmental deviations E_O, E_M and E_F are independent and normally distributed with mean 0 and variance equal to V_E.

Generalizations within this modeling framework may be used to handle situations with environmental correlations between relatives as well as some of the other complications discussed in section 4.1.2. To do this, we consider E as an environmental value ascribed to the individual and include the possibility of "transmission" of environmental values, that is, the environmental value of the offspring may be influenced by the environmental values of the parents and by their phenotypic values:

$$E_O = \frac{\eta(E_M + E_F)}{2} + \frac{\beta(P_M + P_F)}{2} + y,$$

where η and β are coefficients that describe the influence of the parental values on E_O. Here y is a normally distributed random variable with mean 0 and a variance that describes the influence of the offspring's actual environment on E_O where E_O deviates from its mean value given by the parents. The more general model is now fully specified by writing the fundamental equation that describes the phenotype of the individual as a consequence of genetic and environmental influences:

$$P = \Gamma + E + z.$$

Here z is a random variable with mean 0 and a variance that describes the influence of environmental "noise" on the determination of an individual's phenotypic value from its genetic and environmental values.

This more general model produces new expressions for the covariances among relatives. The covariances of Table 4-1 are the covariances of the genotypic values, Γ, among relatives, but now the covariances of the environmental values, E, among relatives must be included. However, these covariances will in general change between generations in the frequency model. Therefore, in the application of the model it is usually assumed that the population has reached an equilibrium, so that characteristic values for the familial covariances can be found.

The present modeling framework was introduced by Sewall Wright in 1931, and for its analysis he developed a statistical technique, called *path analysis* (see, for example, Li, 1973). The method has been extended and used in, for instance, the interpretation of familial aggregation of chronic diseases (Sing and Skolnick, 1979), and many other genetical applications. Outside genetics, it has been widely used in econometrics and sociology.

Extensions of these path analysis models are numerous. Assortative mating can be allowed as in the models of Wright (1931) and Rao and colleagues (1976). Similar models of Feldman and Cavalli-Sforza (1979b), however, invoke the effect of assortative mating differently, and produce quite different results in some instances. Another extension, due to Morton and Maclean (1974), allows the analysis of the effect of a major locus on the character. These models are still under discussion since various ways of introducing simplifying assumptions are possible (Cavalli-Sforza and Feldman, 1978; Feldman and Cavalli-Sforza, 1979b; Goldberger, 1978a,b). It should be stressed that the linear way in which the variables are introduced is essential to the method of analysis and may not always be a legitimate representation of reality. Furthermore, there may be statistical difficulties in comparing estimated values of β and η with expectations. A final point is that the inclusion of more linear components, or different types of covariances among the components, may well produce qualitatively different estimates of the coefficients. This is illustrated by the change in the estimate of heritability of IQ made in 1976, 68 percent (Rao et al., 1982) to one of about 35 percent in 1982 (Rice et al., 1980; Rao et al., 1982) even though both used path analysis techniques.

4.1.5 Exploratory Major Gene Analysis

In recent years attempts have been made to interpret familial aggregation in ways that do not rely entirely on the linear framework typical of polygenic inheritance models (see section 1.1.3.2).

A unimodal trait distribution does not preclude the possibility that a large fraction of the phenotypic variance is due to the genotypic variance at a single locus. Using the data in Figure 1-30, 66 percent of the variance of the distribution of red cell acid phosphatase activity, shown in Figure 1-4, can be explained by genotypic variation at the electrophoretically defined red cell acid phosphatase locus (section 4.1.1). Assuming random mating, about 15 percent of all matings in the population (Table 1-13) will be between individuals of the same homozygous genotype at the electrophoretic locus. Thus, the variance among sibs in these families will be at most 33 percent of the phenotypic variance. On the other hand, sibs from heterozygote × heterozygote matings will exhibit a variance close to the phenotypic variance in the population. This means that the pattern of familial aggregation will be strongly influenced by the segregation of a gene with large effects on the character. Such a gene is called a *major gene,* and as we have just seen, its presence is expected to cause heterogeneity in the variance among sibs between families. This heterogeneity is usually the only clue to the existence of a major gene since it is the familial aggregation of *phenotypic* variation that is observed.

Approaches to the detection and analysis of major gene effects vary with the character under study and the type of data available. One approach considers the offspring of a family in which it is assumed that genes affecting the phenotype under study are segregating. The variance of the trait among these sibs should vary more as a function of the average value among the sibs if there is a major gene than if the trait is polygenic. In addition, the variance in certain sibships could be higher than the phenotypic variance in the population. The relationship expected between the within-sibship variance and mean may be investigated using computer simulation of one-locus, two-allele models. The allele frequency p_1 and mean values Γ_{11}, Γ_{12}, Γ_{22} of the phenotype for genotypes A_1A_1, A_1A_2, A_2A_2 are the parameters of the simulation model (see section 4.1.1); in addition, a value of the variance of the phenotype for each genotype (assumed equal for all genotypes) is specified.

Parental combinations producing the sibships are chosen according to the Hardy-Weinberg proportions, and the offspring are also chosen at random using the Mendelian segregation rates (see Exercise 2.2.C). The result is a set of computer-generated data that can be used to test whether the variance within sibships does indeed vary with the mean among the sibs. Fain (1978) showed that this dependence could easily be detected in simulated data. He then took sets of familial data on height, weight, and the results of a test of spatial ability (a mental rotations test) and found that only the last actually showed the statistical increase of variance with mean, among sibs, that the simulation predicted.

This exploratory approach has been substantially extended by Karlin and colleagues (see, e.g., Karlin et al., 1981 and Karlin and Williams, 1981). These extensions use the same simulation approach as described earlier, but apply it to describing the properties of a wide array of functions of the difference in phenotypic value between offspring and parents. The behavior of each of these functions is characteristic of whether the simulated model is polygenic or a major gene model. These functions, or indicies, are then computed from observed family data and the values compared with those obtained under simulation. From this comparison it is suggested, for example, that the pattern of familial aggregation of plasma triglycerides is well explained as the consequence of a major gene. The use of these functions, whose properties are relatively model free, and whose expected values can be ascertained by a combination of mathematical and computer simulation techniques, is called *structured exploratory data analysis*.

Detailed description of these techniques is beyond the scope of this book. Suffice it to say that exploratory methods are under active development. No single technique is regarded as sufficient to detect a major gene or to distinguish between major gene and polygenic models. For this a combination of approaches, including the linear models of section 4.1.4, will most likely be needed (see also section 4.2.1). It must be realized, however, that the description of the familial aggregation in terms of major gene models suffers from the same weakness as the polygenic models, namely, that the genetic variation cannot be monitored directly. For instance, the environmental effects on familial aggregation, considered to be continuous in section 4.1.3, may equally well be discrete and simulate major gene effects.

4.2 Discrete Disease Phenotypes and Heritability of Liability

Falconer (1965) suggested a method that uses Fisher's additive genetic model of section 4.1 to estimate the *heritability of the liability* assuming a normally distributed liability to a disease (section 1.1.3): The liability is a theoretical character, a sufficiently high value of which causes the disease to be manifest. This cutoff value is called a *threshold* (Figure 4-5). Individuals whose liability falls above the threshold T have the disease, and among the general population these comprise a fraction d_g, the disease rate. The mean liability in the population is supposed to be μ_g, while that among the affected individuals is μ_d. Here we use the Greek letter μ to distinguish these theoretical means from observable means. The relatives of an affected individual will, due to familial correlation of the liability trait, have a distribution of liability values with a mean closer to the threshold, as illustrated by the liability distribution in Figure 4-5. Among these relatives the mean liability is supposed to be μ_r and the fraction with the disease is d_r, as deter-

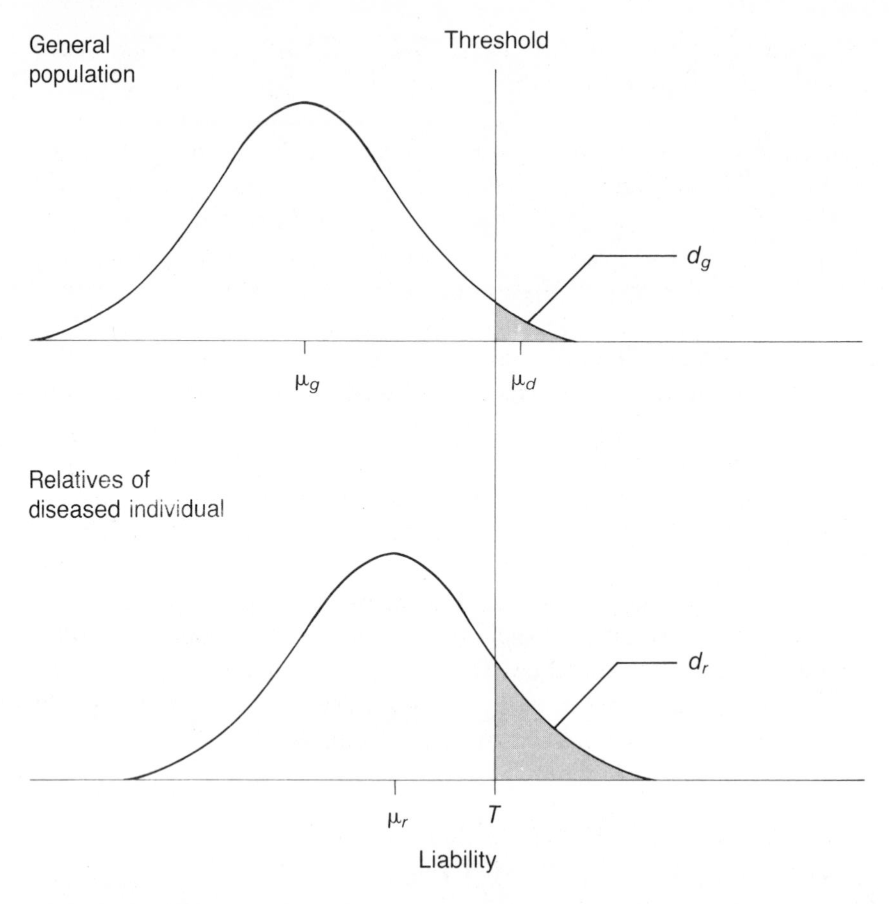

Figure 4-5. The liability model as used to describe the familial aggregation of multifactorial diseases.

mined from familial aggregation studies. Usually, the considered relatives are close relatives of the probands, that is, their offspring, parents, or sibs.

Under the assumption that the liability is an additive character, transmitted according to the model of section 4.1.1 with $V_G = V_A$ ($V_D = 0$), it is known that for the considered relatives (Figure 4-5)

$$\frac{\mu_r - \mu_g}{\mu_d - \mu_g} = h^2\omega,$$

where the coefficient ω of the heritability h^2 is determined from the degree of relationship between the affected individuals and their rel-

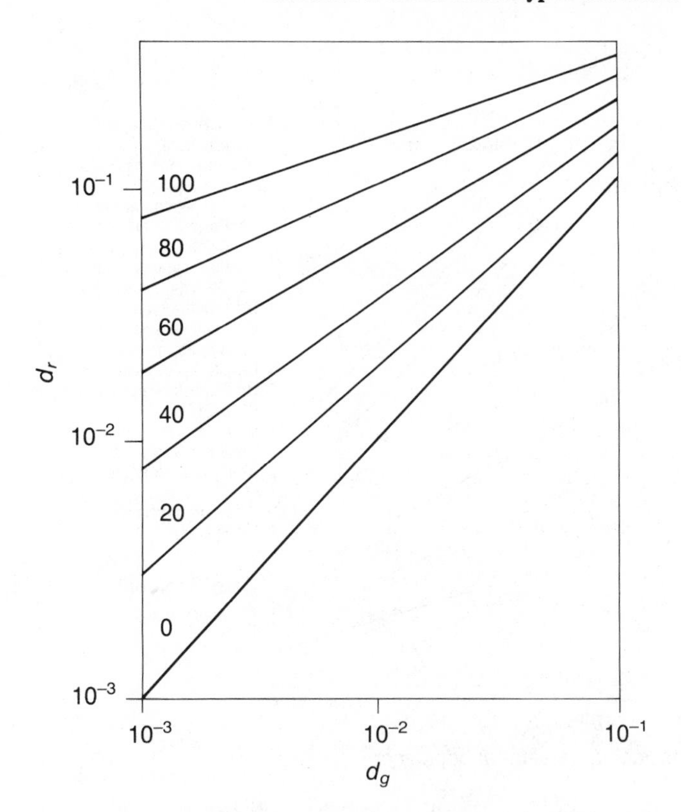

Figure 4-6. Heritability of liability (percent) in terms of d_g*, the disease incidence in the general population, and* d_r*, the disease incidence among relatives of affected individuals. The table applies for* $\omega = 1/2$*, i.e., for sibs or between parent and offspring. (After Falconer, 1965).*

atives: for example, for the parent-offspring or sib relationships, $\omega = 1/2$. From analytical properties of the normal distribution, the expression for the heritability can be rearranged. In particular, the numerator can be written in terms of the threshold, T, as

$$\mu_r - \mu_g = (T - \mu_g) - (T - \mu_r),$$

and the values of $T - \mu_g$ and $T - \mu_r$ bear a simple relationship to the observable disease rates d_g and d_r, respectively. Figure 4-6 presents a useful representation of the heritability, h^2, as a function of d_g and d_r in close relatives.

A general survey of a number of diseases in terms of their incidence in the Caucasian population and among sibs is given in Figure 4-7. In

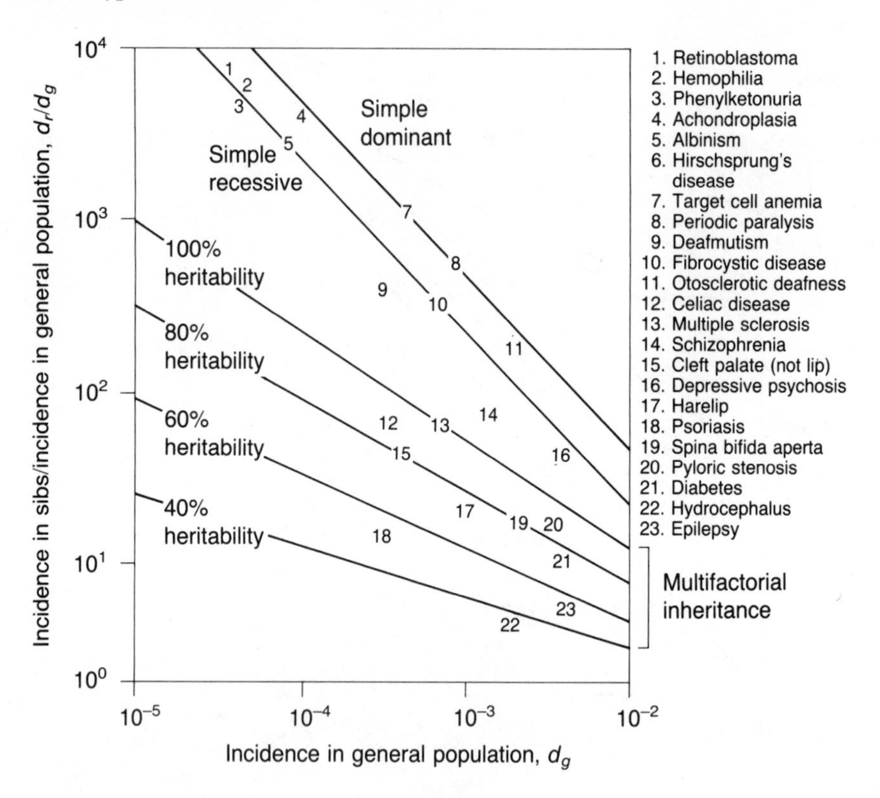

Figure 4-7. Observed relations between disease incidences in the general population and among sibs of affected individuals. The relation is given as d_g *to* d_r/d_g *for a number of diseases. Heritabilities of liability for multifactorial diseases are taken from Figure 4-6. and the expected relations for Mendelian diseases are computed simply as* $d_g/4$ *for recessive diseases and* $d_g/2$ *for rare dominant diseases. (After Cavalli-Sforza and Bodmer, 1971.)*

evaluating the heritabilities of liability in this figure, remember that no liability function has yet been empirically determined.

The use of the same hypothetical liability function in the relatives of the diseased as in the general population has been questioned. Genes that contribute to the liability are expected to be more frequent among relatives of probands than in the general population. Further, the disease may have variable age of onset, so the sample of relatives, as well as the sample of the general population, is likely to be heterogenous for incidence of the disease simply because of its age structure (making the standardization referred to in Chapter 1 essential). In addition, the threshold may vary with the environment, in which case standardization for socioeconomic, dietary, or cultural contributions to the disease may be important.

Indirect evidence for the existence of a liability trait underlying a qualitative character is obtained from the experiments with *Drosophila melanogaster* by Waddington (1953). Treatment of *Drosophila* pupae by heat shock caused about a third of the adults to exhibit an aberration in wing morphology, which was otherwise extremely rare in the population. Selection for this trait in treated flies produced an increasing incidence of the trait after heat shock treatment. However, an increasing fraction of the flies concurrently showed the trait spontaneously without the treatment. This finding is explained by the liability model if it is postulated that the heat shock treatment changes only the threshold of expression of the trait (Figure 4-8).

4.2.1 Pedigree Analysis of Complex Transmission

Occasionally the investigator is fortunate enough to obtain one or more large pedigrees with an aggregation of a disease. This allows a more detailed investigation of the mode of transmission of the disease in terms of the probability that a certain individual will be affected by the disease. The latter does not, in principle, require knowledge about the mode of transmission because it might be estimated on empirical grounds alone. However, a reliable genetic description of the transmission will provide more accurate and generally applicable estimates.

Elandt-Johnson (1970) has pioneered the analysis of transmission rules in what has become known as the *pedigree analysis* (see, for example, Elston, 1979, for a review). One way to specify the transmission model in its simplest form is to assume that the disease is determined by a single locus with two alleles A_1 and A_2. It is unknown whether an individual in the pedigree is one of three genotypes A_1A_1, A_1A_2, A_2A_2, but the idea of the analysis is to choose from among a set of alternative genetic rules (dominance, recessivity, X-linkage, etc.) the rule that makes the observed pedigree most likely. For example, a model might assume complete dominance so that affected individuals are A_1A_1 or A_1A_2. Usually a nongenetic "transmission" rule in which each individual, affected or not, transmits the disease at the same rate to an offspring is taken as an alternative mode of transmission to these genetic rules. This purely "environmental" alternative is used for comparison to assess whether the pedigrees contain any evidence that the probability of acquiring the disease is influenced by the genotype of the individual.

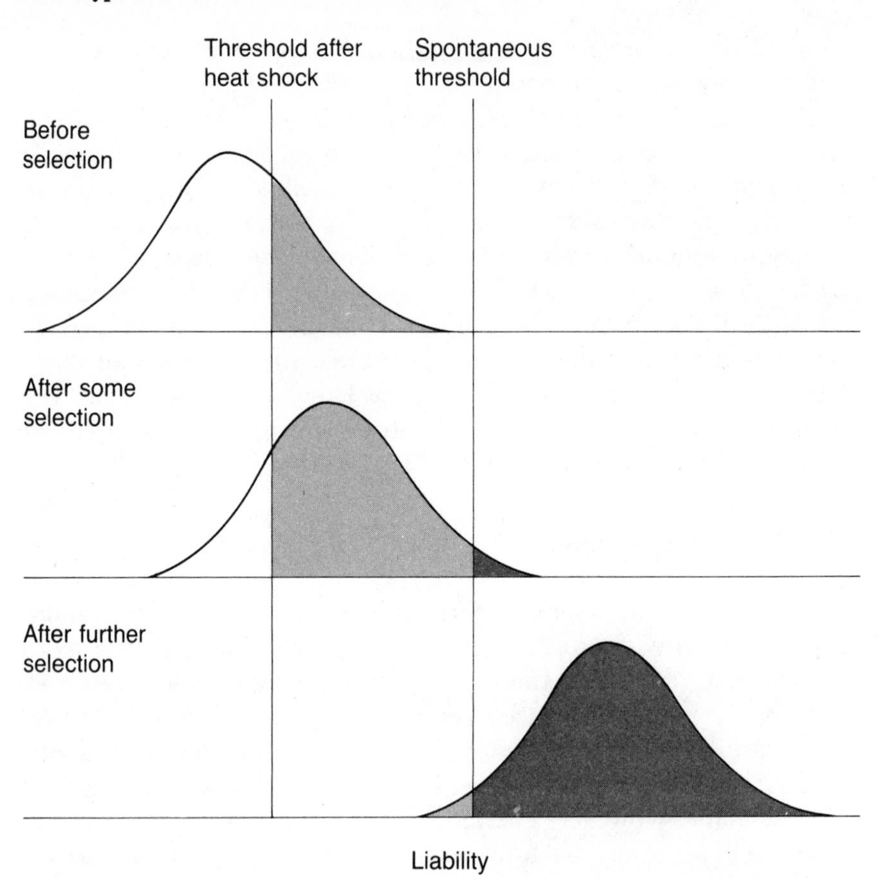

Figure 4-8. Sketch of the interpretation of Waddington's experiment in terms of the liability model (Falconer, 1965). The description of liability in Figure 1-13 applies, but the norm of reaction of the individuals to an increased temperature is assumed to result in a lowering of the threshold. Selection for flies with abnormal wings after heat shock of the pupae then amounts to selection for flies with high liability. Selection for high values of a heritable quantitative character usually produces a population with an increasing mean value for the character. Thus the prediction is that the mean liability increases under selection, with the result that the rate of both spontaneous and induced aberrations in wing morphology increases.

King and colleagues (1980) examined 11 pedigrees in which breast cancer seemed to aggregate. With a one-locus, two-allele model, the most likely choice turned out to be autosomal dominant. In a larger study of 200 Danish pedigrees, Williams and Anderson (1984) reached the same conclusion. They showed that a model with a single dominant autosomal locus could explain all of the transmissible variation. Further, they found a high transmission rate for cancers with an early age of onset, but that those with a late age of onset were considerably less transmissible (Figure 4-9). The limited number of choices used in such treatments must be a cause for some skepticism (see, for example, Feldman and Cavalli-Sforza, 1979a), and it would be desirable to have other corroborating evidence before counseling is based on the acceptable fit to such a model. Of course, a set of pedigrees should not be viewed as representing familial aggregation in the population without more supporting evidence.

The principles of pedigree analysis have been widely used to estimate the *recombination fraction* between two markers. This is done by computing the probability $P(R)$ of the observed pedigree given a recombination fraction R and comparing that to $P(\frac{1}{2})$, the probability

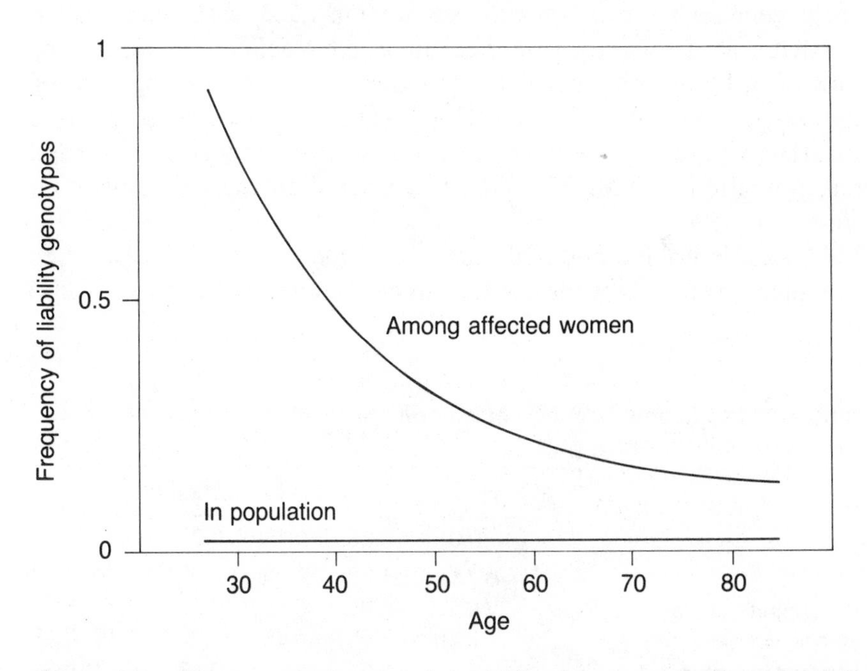

Figure 4-9. Heterogeneity in transmission of breast cancer according to age of onset. The familial aggregation and transmission of breast cancer is described by the segregation of a rare dominant autosomal allele (a liability allele) with a frequency in the population of 0.008. A large proportion of young women with breast cancer carry the liability allele. This fraction decreases steadily with age. Using the liability model, the probability of acquiring the disease is estimated to be more than 10 times higher for women carrying the liability allele than for those carrying the recessive allele. (After Williams and Anderson, 1984.)

given the genes are unlinked $\left(R = \frac{1}{2}\right)$. The value

$$z(R) = \log_{10}\left[\frac{P(R)}{P(1/2)}\right]$$

is computed for each available pedigree and the sum of $z(R)$ values over the pedigrees is called the *lod* (log odds) *score* corresponding to R. If the value of z is greater than 3 for a small value of R, say, less than 0.1, then the pedigree set is more than 1000 times as likely to have occurred with $R \leqslant 0.1$ than with $R = \frac{1}{2}$. In other words, it is reasonable to conclude that the genes are linked. It is usual to compute the lod score for recombination values from 0 to 0.4 in order to obtain a feeling for the most likely value of the true recombination fraction. Table 4-2 presents a recent use of this technique to show the linkage of the gene for Huntington's chorea (section 1.2.2) with a gene called *G8* that is actually a segment of chromosome 4 defined, using a probe called *G8*, by two polymorphic restriction sites that are recognized by the restriction enzyme HindIII (Gusella et al., 1983). The table shows that the two genes are very tightly linked while neither show restricted recombination with *MNS* or *GC*, two other allelic polymorphisms on chromosome 4.

This statistical method of determining linkage has been used (in an extended form to allow for one trait to have complicated transmission)

Table 4-2. *Lod Scores for the G8 Marker with Huntington's Chorea and other Chromosome 4 Markers.*

	Recombination fraction (θ)					
	0.0	0.05	0.1	0.2	0.3	0.4
	A 1.81	1.59	1.36	0.90	0.48	0.16
Huntington's chorea against *G8*	V 6.72	5.96	5.16	3.46	1.71	0.33
	T 8.53	7.55	6.52	4.36	2.19	0.49
Huntington's chorea against *MNS*	− ∞	−3.22	−1.70	−0.43	−0.01	0.07
Huntington's chorea against *GC*	− ∞	−2.27	−1.20	−0.32	0.00	0.07
G8 against *MNS*	− ∞	−8.38	−3.97	−0.55	0.45	0.37
G8 against *GC*	− ∞	−2.73	−1.17	−0.08	0.14	0.08

From Gusella et al. (1983).
A, American pedigree; V, Venezuelan pedigree; T, total.

for example to establish the linkage between HLA and hemochromatosis (Kravitz et al., 1979). When the genetic map location of a disease gene is not known as, for example, with the muscular dystrophies, close linkage to a polymorphic DNA segment can be of substantial diagnostic assistance. DNA polymorphisms have been shown to be associated with fragile X–mental retardation syndrome, Huntington's chorea (see earlier), muscular dystrophy (Becker or Duchenne), and others (Cooper and Schmidtke, 1984). In general, linkage has the potentially important role of confirming or at least supporting the suspicion that a trait is influenced by the genotype. In addition, in the absence of biochemical ways of distinguishing different forms of a disease it may be possible to classify them on the basis of linkage to one or more markers. Such linkage may, as will be mentioned later, be an important aid in genetic counseling.

4.3 Phenotypic Variation in Twins

In humans, monozygous (MZ) twins are the only individuals with completely equivalent genomes and for this reason they are a valued subject for study. There are racial differences in the frequency and nature of twinning. For instance, African blacks have a higher frequency of twinning than Caucasians, with a substantial bias toward dizygous (DZ) twins. Orientals have fewer twins than Caucasians, but most are monozygous. As with any special phenomenon in biology, the goal in studying twins is to extrapolate to more usual population situations. This is not straightforward, as the special relationship between twins, especially monozygous twins, has become almost legendary. It may even be the case that in families with twins and other children, the twin-nontwin relations might also be special (see, for example, Kamin, 1974). For a detailed review of the biology and what is known of the familial aggregation of twinning itself, the reader is referred to Bulmer (1970).

For most morphometric traits, the differences between pairs of monozygous twins are significantly less than those between dizygous twin pairs. These differences are expressed by the variance W within twin pairs for each type of twins, that is, W_{MZ} and W_{DZ}. These in turn are estimated by the average of the squares of the differences between monozygous twins ($2W_{MZ}$) and between dizygous twins ($2W_{DZ}$), the average being taken over all twins of the respective types. If the variance of the environmental effects on the phenotype of the twins does not depend on zygosity, then W_{MZ} should estimate the proportion of W_{DZ} due only to environmental variation, as the variance between monozygous twins is due entirely to the environment. Thus, under the assumption of equal environmental effects, the quantity

$$H = \frac{W_{DZ} - W_{MZ}}{W_{DZ}}$$

estimates the fraction of dizygous twin variance due to genetic variation; some examples are given in Table 4-3.

Another similar method used in the comparison of monozygous and dizygous twins uses the intraclass correlation coefficient ρ between pairs of a given type of twins. For a given collection of pairs of individuals,

$$\rho = 1 - \frac{W}{V'},$$

where W is half of the average squared difference between the members of a pair and V' is the total variance of all the observations of a given zygosity (see section 1.1.3.1). Using the subscripts MZ and DZ to denote values taken from collections of these respective sets of twins, and assuming $V'_{MZ} = V'_{DZ}$, the expression

$$H' = \frac{\rho_{MZ} - \rho_{DZ}}{1 - \rho_{DZ}}$$

Table 4-3. *Estimates of H from Monozygous and Dizygous Twin Data.*

Trait	Male	Female
Stature	0.79	0.92
Weight	0.05	0.42
Arm length	0.80	0.87
Foot length	0.83	0.81
Hip circumference	0.19	0.66
Cephalic index (head breadth/head length)	0.90	0.70
Masculinity-femininity index	0.78	0.85

Data from Osborne and DeGeorge (1959).

Table 4-4. *Genetics of Cleft Lip and Palate.* Number of concordant pairs of twins among 125 monozygous and 236 dizygous twin pairs where at least one of the twins had cleft lip and palate. The level of concordance is obviously much higher in monozygous than in dizygous twins.

Twins	Concordant pairs	Discordant pairs
MZ	37	88
DZ	11	225

After Vogel and Motulsky (1979).

should closely approximate H (see Cavalli-Sforza and Bodmer, 1971, ch. 9).

Monozygous twin pairs in which one or both are reared from birth by foster parents that are chosen *randomly* from the general population should, in principle, provide the perfect experiment to determine the degree of genetic versus environmental determination of a trait. The two largest reputable studies of separated monozygous twins, those of Shields (1962) and Newman, Freeman, and Holzinger (1937), total 63 pairs, of whom about half were raised in the families of biological relatives. Random adoption is now impossible in most civilized societies, and the statistical problems concomitant to nonrandom adoption are profound.

Studies of adopted children who are not twins can also contribute to our feeling for the level of environmental determination of a trait. Such studies were pioneered by Burks (1928) and Leahy (1935), who found that for IQ the adopted child was generally more strongly correlated with its true parent than with foster parents. Some important caveats to these findings have been pointed out by Kamin (1974). For example, problems arise from nonrandom adoption because of the attempts by adoption agencies to match the foster and biological parents as closely as possible.

Studies of concordance of monozygous pairs and dizygous pairs for a polygenic, discretely defined disease give a useful indication of the degree to which the disease is genetic. Association tables such as Table 4-4 can be constructed, and the ratio of monozygous to dizygous twin concordances can be evaluated and tested for significance. The significant difference between the concordances does not, however, produce specific etiological information; it suggests only that there is some genetic factor involved.

5

Mendelian Genetic Diseases in Populations

The description of Mendelian genetic diseases given in section 1.2.2 suggests that directional selection should occur against the disease allele. The conclusion from section 3.2.1 is, then, that the population should become monomorphic for the normal allele. This conclusion, however, ignores the role of mutation. In the following we develop the theory for the population distribution of Mendelian diseases as a balance between the production by mutation of the disease alleles and the elimination of these alleles by selection against diseased individuals. This theory, and indeed population genetics theory in general, assists in the understanding of the dynamics of the population incidence of these diseases. In turn it may prove valuable in the counseling of families in which a disease is observed and in the evaluation of the general effects of clinical procedures for specific diseases.

5.1 Population Genetics of Mendelian Diseases

Mendelian genetic diseases occur as a result of mutation from the normal allele to the disease allele, which occurs at the mutation rate, μ. The disease allele is usually rare, so that back mutation at rate μ will not be of any importance in the determination of the disease rate. We will therefore assume in the following that the back mutation rate is zero. Further, the diseases are termed recessive and dominant according to their fitnesses. A gene with the normal allele a and the disease allele A causes a dominant disease if the fitnesses of the genotypes aa, Aa, and AA are 1, $1 - s$, and $1 - t$ with $0 < s < t$. A gene with the normal allele B and the disease allele b causes a recessive disease when BB, Bb, and bb have the fitnesses 1, 1, $1 - t$ with $0 < t$.

5.1.1 Mutation-selection Balance

First consider an autosomal dominant disease. Let the frequency of allele a be p ($q = 1 - p$ is the frequency of allele A) among zygotes in a given generation. Then, with random mating, the allele frequency of a among breeders is $p^\star$, which can be calculated from section 3.2.1 as

$$p^\star = \frac{p(1 - sq)}{1 - 2spq - tq^2},$$

where the relative viabilities of aa, Aa, AA are $v_{aa} = 1$, $v_{Aa} = 1 - s$, $v_{AA} = 1 - t$. When these individuals form gametes, mutation in the germ line cells become apparent, and the allele frequency p' of a among offspring can be calculated from section 3.1.1 as

$$p' = (1 - \mu)p^\star.$$

The change in allele frequency may then be seen in Figure 5.1, which is a hybrid between Figures 3.1 and 3.3. When p is small, the figure is indistinguishable from Figure 3.3 and p increases due to directional selection, but when p approaches 1, mutation balances selection at a globally stable mutation-selection balance equilibrium $\hat{p}$. For reasonable mutation rates, $\hat{p}$ is very close to 1, and instead of finding $\hat{p}$ as the point of intersection between the selection curve and the mutation line, we may approximate the selection curve by its tangent at $P = 1$ (Figure 5.2). With this approximation the mutation-selection balance equilibrium for a *dominant* disease becomes

$$\hat{p} = 1 - \frac{\mu}{s} \text{ and } \hat{q} = \frac{\mu}{s},$$

which corresponds to the outcome obtained by neglecting the occurrence of the extremely rare genotype AA.

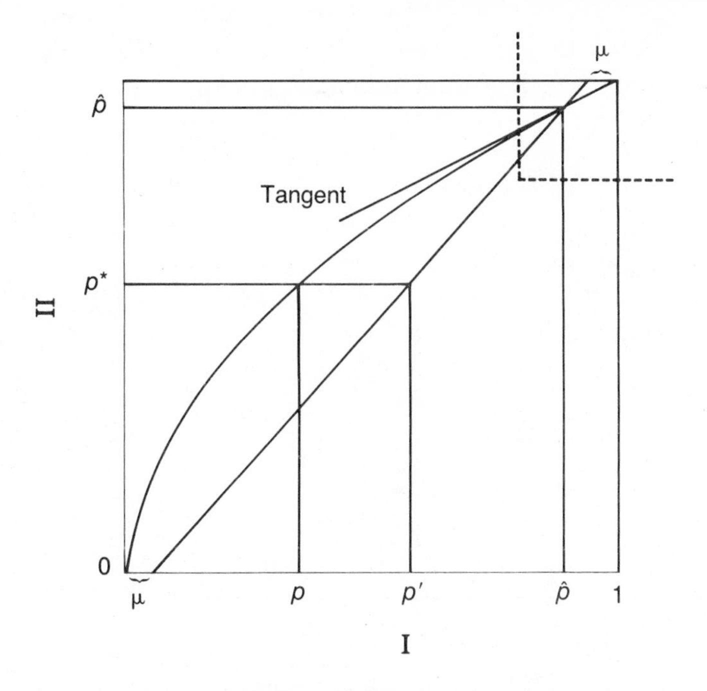

Figure 5-1. The change in allele frequency between generations due to selection and mutation ($v_{AA} = 0.1$, $v_{Aa} = 0.5$, $v_{aa} = 1$)*. The parental allele frequency of* a, p, *on axis I produces the allele frequency* p★ *after selection (see Figure 3-3), and the allele frequency* p★ *on axis II produces the allele frequency* p′ *among zygotes after mutation (the mirror image of Figure 3-1). Mutation rates (from* A *to* a *and* a *to* A*) shown here are unrealistically large.*

For autosomal recessive diseases this approach does not work (Figure 5.3), because to neglect the occurrence of *bb* is to neglect the occurrence of the disease. However, the change in allele frequency p of the normal allele B is given by the foregoing equations for $p^\star$ and p', but now with $s = 0$. In this case, solution of the equation $p' = p$ produces the *recessive* mutation-selection balance equilibrium as

$$p = 1 - \left(\frac{\mu}{t}\right)^{1/2} \text{ and } q = \left(\frac{\mu}{t}\right)^{1/2}.$$

Dominant diseases with s very small are almost recessive, so this approximation to the equilibrium for a dominant disease becomes invalid. Indeed, if s is less than about μ, then the recessive mutation-

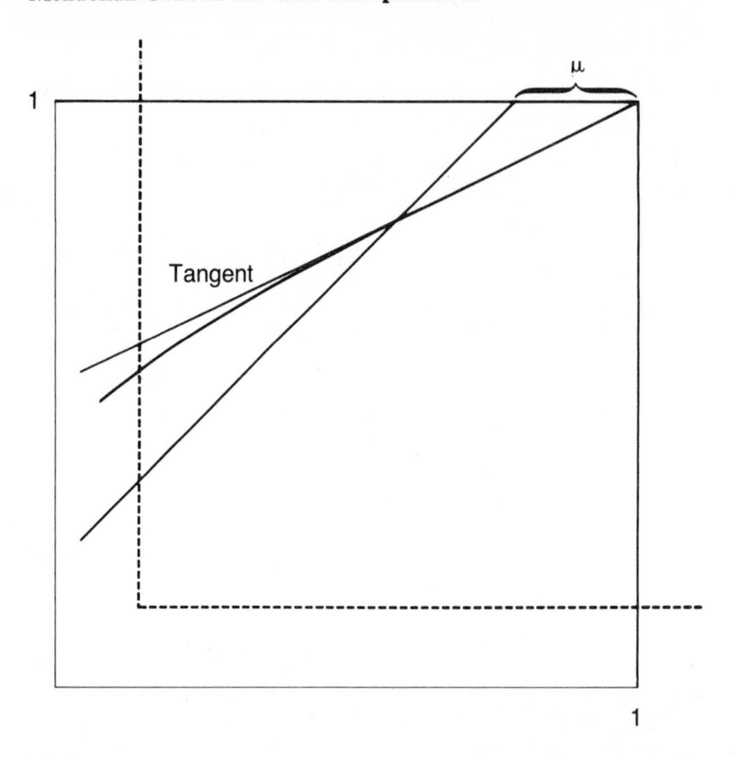

Figure 5-2. Magnification of the upper right corner of Figure 5-1.

selection balance equilibrium is a better approximation than the equilibrium for dominant diseases as shown in Figure 5.4.

Mutation-selection balance for X-linked genes is similar to that for dominant diseases in autosomal genes, since a disease allele will always be "dominant" in the males even if it is recessive in females. Suppose the genotypic fitnesses in the females are 1, $1 - s$, and $1 - t$ for the genotypes BB, Bb, and bb, and 1 and $1 - t$ for the male genotypes B and b. At equilibrium we have approximately

$$\hat{q}_{♀} = \frac{\mu(3 - s)}{2s + t - st}$$

$$\hat{q}_{♂} = \frac{\mu(3 + s - 2t)}{2s + t - st},$$

for a disease that is dominant ($0 < s < t$) in the females, and

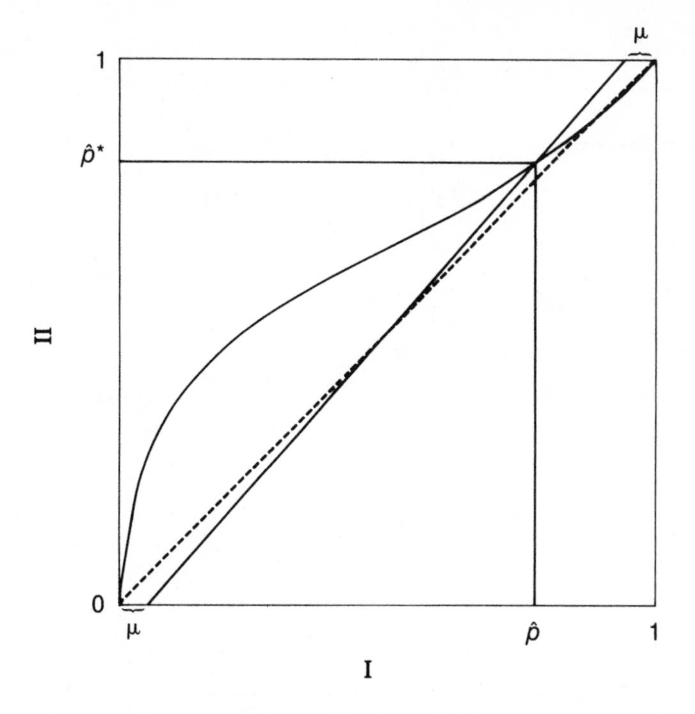

Figure 5-3. The change in allele frequency for a recessive disease between generations due to selection and mutation ($v_{BB} = v_{BB} = 1$, $v_{bb} = 0.1$). *Otherwise as for Figure 5-1.*

$$\hat{q}_{♀} = \frac{3\mu}{t}$$

$$\hat{q}_{♂} = \frac{\mu(3 - 2t)}{t}$$

for a recessive disease ($s = 0$).

Mutation-selection balance equilibria for diseases determined by two or more loci have not been considered extensively, although some theory was developed by Christiansen and Frydenberg (1977). The case of favism and *G6PD* deficiency may be an example of such a two-gene mutation-selection system (see Cavalli-Sforza and Bodmer, 1971).

5.1.2 Disease Frequencies and Allele Frequencies

For a dominant disease the frequency of the disease genotype is about $2\mu/s$, and for a recessive disease it is μ/t. Thus, the frequency of the

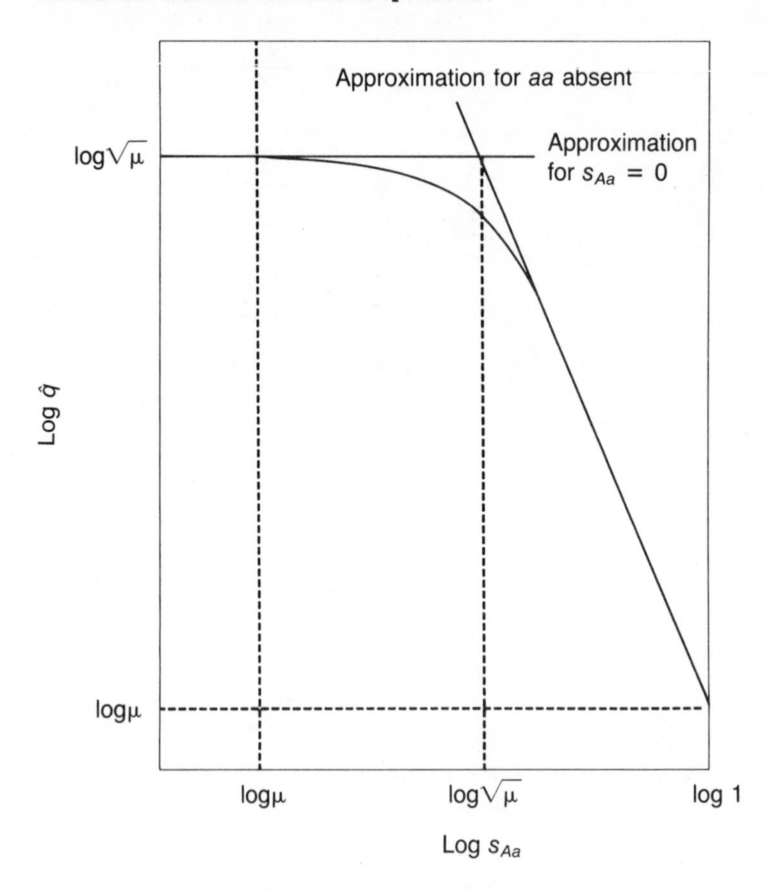

Figure 5-4. The equilibrium frequency of the disease allele as a function of the selection coefficient of the heterozygote, $s_{Aa} = 1 - v_{Aa}$, *in the situation where* $s_{AA} = 0$ *and* $s_{aa} = 1$, *that is,* $v_{AA} = 1$. $v_{aa} = 0$. *The approximations used for dominant* (aa *absent) and recessive* ($s_{Aa} = 0$) *diseases are shown.*

disease is of the order of the mutation rate for both types of diseases. The gene frequencies of the two types are of quite different orders of magnitude, however, and the population dynamics of the diseases are qualitatively quite different.

The basis of the mutation-selection balance can be expressed in terms of an approximate balance equation. The production of disease alleles per generation is about μ, the loss due to selection is about qs for dominant diseases and about q^2t for recessive diseases. At equilibrium we have for the disease allele:

Production by mutation = loss by selection.

Thus for a dominant disease a fraction $1 - s$ of the individuals with the disease genotype have inherited the disease allele from their parents and a fraction s are new mutants. For a recessive disease only a fraction $2(\mu t)^{1/2}$ of the individuals with the disease genotype carry a newly mutated gene, and most affected individuals have inherited both disease genes from their parents. Thus the dominant disease alleles are expressed immediately and eliminated rather soon after mutation, whereas the recessive mutations are hidden in a pool of heterozygous carriers of the disease. Only a tiny fraction of all the recessive disease genes is expressed in the population. We can expect, therefore, that any change in the environment of the population should immediately result in changed disease frequencies for dominants, whereas change in the rate of occurrence of recessive diseases should occur very slowly (section 3.2.2).

As remarked in section 3.2.5, the average fitness of a rare recessive allele, $1 - tq$, deviates only slightly from the average fitness of the normal allele. Indeed, at equilibrium this average fitness is $1 - (\mu t)^{1/2}$ compared to the average fitness of the population of $1 - \mu$. Thus, the allele frequency change in a population close to equilibrium is comparable to that in the case of mutation-mutation balance (section 3.1.1). Genetic drift is therefore likely to play a role in determining the actual allele frequency for a recessive disease. Figure 5.5 shows the distribution of allele frequencies for a recessive lethal disease ($t = 1$) in a large number of populations of a given size. As the figure shows, the gene frequencies only cluster around the equilibrium $\hat{q} = \mu^{1/2}$ if the number of mutations per generation, $2N\mu$, is bigger than about 1. Differences between populations in the incidence of recessive diseases are therefore expected since few human populations have had an effective population size above 10^5–10^6 for an extended period of time.

Exercise 5.1.A

The recessive autosomal disease cystic fibrosis of the pancreas occurs at a frequency of 0.0004. The disease is usually fatal before maturity.

a. Assuming random mating, what is the frequency of the disease gene in the population of newborns?
b. If the rate of mutation from the normal allele to the disease allele is 4 ×

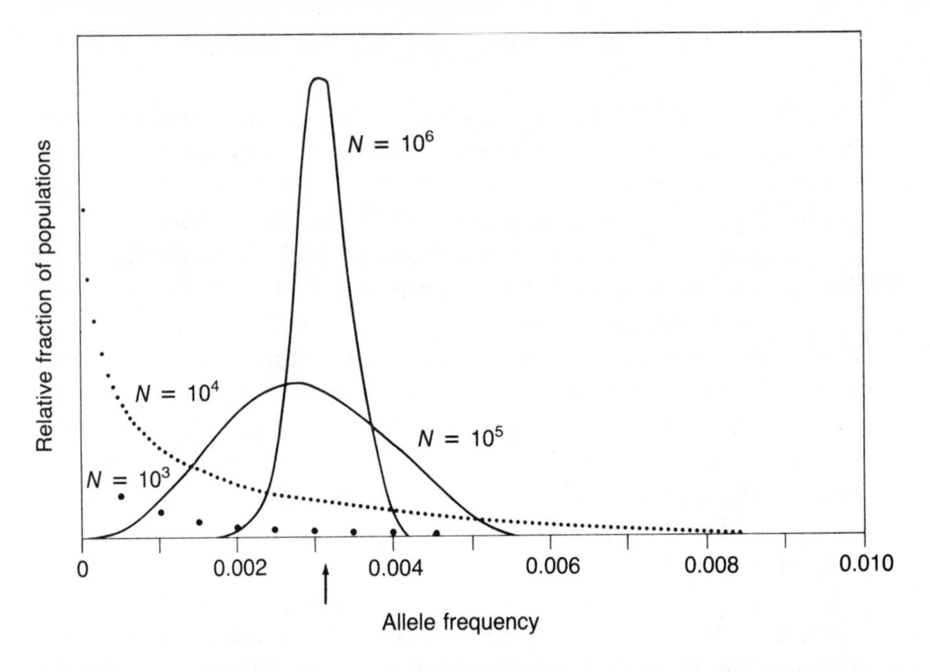

Figure 5-5. The distribution of allele frequencies in a large number of populations of size N *for a recessive lethal* ($v_{BB} = v_{Bb} = 1$, $v_{bb} = 0$). *The mutation frequency is* 10^{-5}, *so* $\hat{q} = 0.0032$, *shown by the arrows. (After Wright, 1937).*

10^{-6} per generation, what is the frequency of the disease allele at the mutation-selection balance equilibrium?

c. Assuming the human generation time to be 30 years, how many years would be required to reduce the frequency of the disease allele from the value in (a) to 0.01?

d. What could be the explanation for the high frequency of cystic fibrosis?

Exercise 5.1.B

In an investigation by Mørch (1941) of about 94,000 newborns, 10 cases of chondrodystrofic dwarfism were found. This condition is known to be caused by an autosomal dominant allele.

a. What is the frequency of this dominant allele in the population?

These dwarfs survive normally, but in an investigation of adult dwarfs 108 individuals had a total of 27 children, whereas 457 of their normal sibs had a total of 582 children.

b. What is the fitness of the dwarfs relative to their normal sibs?

c. Accept this relative fitness as the fitness of dwarfs relative to normals, and use it to estimate the mutation rate from the recessive normal allele to the dominant disease allele, assuming mutation-selection balance equilibrium.

Of the 10 newborn dwarfs, 3 had one parent that was chondrodystrofic, whereas 7 had normal parents.

d. Use this information to calculate the mutation rate.
e. From the information that 3 of the 10 newborn dwarfs had a dwarf parent, estimate the relative fitness of dwarfs to normals and specify the assumptions needed to make this estimate.

5.1.3 Inbreeding and Recessive Diseases

Inbreeding in a population increases the number of homozygotes at any autosomal locus (section 2.4.1). This is especially important in the case of rare recessive Mendelian diseases. Suppose the disease allele has the frequency q in two populations, one randomly mating, the other with a population inbreeding coefficient of F. The disease frequency with random mating is q^2, and with inbreeding it is $q^2 + Fpq$, which can be about F/q times greater than with random mating. Thus the rarer the disease, the larger is the effect of inbreeding.

This comparison of disease frequencies is of limited interest, however, because the equilibrium gene frequency is not expected to be the same in the two populations. From our balance equation, the equilibrium disease frequency is the same and equal to the mutation rate in the two populations, so the allele frequency at equilibrium in the inbred population is found from

$$\mu = [F\hat{q} + (1 - F)\hat{q}^2]t.$$

If the inbreeding is small ($F < \mu/t$), we have, approximately,

$$\hat{q} = \left(\frac{\mu}{t}\right)^{1/2} - \frac{F}{2},$$

while for large inbreeding ($F > \mu/t$),

$$\hat{q} = (\mu/t)/F.$$

Thus for small F the frequency of the disease allele decreases approximately linearly with F, and for large F the frequency of this recessive disease allele becomes comparable to that of a dominant disease.

A change in the population inbreeding coefficient changes the disease frequency immediately, but at the same time the globally stable equilibrium changes and the allele frequency moves slowly towards this equilibrium. An equilibrium population with inbreeding F that changes its inbreeding to F' will change the disease incidence by about $(F' - F)$pq and its gene frequency by the same amount in the first generation after the change. If the inbreeding level is small, however, the relative change in disease incidence is far greater than the relative change in the allele frequency. Inbreeding levels in human populations vary through time (Table 2-6 and further data in Cavalli-Sforza and Bodmer, 1971) and the convergence towards equilibrium is very slow (section 3.3.2), so there is no reason to believe that any human population is at allele frequency equilibrium with respect to current or historic inbreeding levels. In recent times the inbreeding level has typically decreased, and if this trend continues, the frequency of recessive disease alleles, and therefore of disease incidence, is expected to increase during the coming millennia.

Recessive diseases are expected to be more frequent among offspring of consanguinous matings. To evaluate this effect, suppose the inbreeding is due entirely to a low frequency u of first-cousin marriages ($f = 1/16$), giving a population inbreeding coefficient of $F = u/16$. The incidence of the disease among children from the inbred marriages is $q/16 + q^2(15/16)$, which is considerably higher than a typical incidence among non-inbred families, q^2. This difference becomes more pronounced the rarer the disease. For instance, a lethal disease ($t = 1$) in a population with $u = 0.01$ and $\mu = 10^{-4}$ will have $\hat{q} = 0.97 \times 10^{-2}$, and 7 percent of the diseased children occur in the 1 percent of consanguinous marriages; for $\mu = 10^{-6}$, we have $\hat{q} = 0.74 \times 10^{-3}$, and 47 percent of the diseased children are inbred. The increased risk for the consanguinous couple may be rather small, but at the population level consanguinity is an important contributor to the frequency of rare Mendelian recessive diseases (Figure 5-6).

Exercise 5.1.C

An autosomal recessive disease occurs in a frequency of 1 in 1,000,000. The disease is lethal in early childhood. Assume random mating and

a. Provide an estimate of the frequency of the disease allele in the population.

b. Guess the rate of mutation from the normal allele to the disease allele.

c. Specify and discuss possible sources of error in your answers to questions (a) and (b).

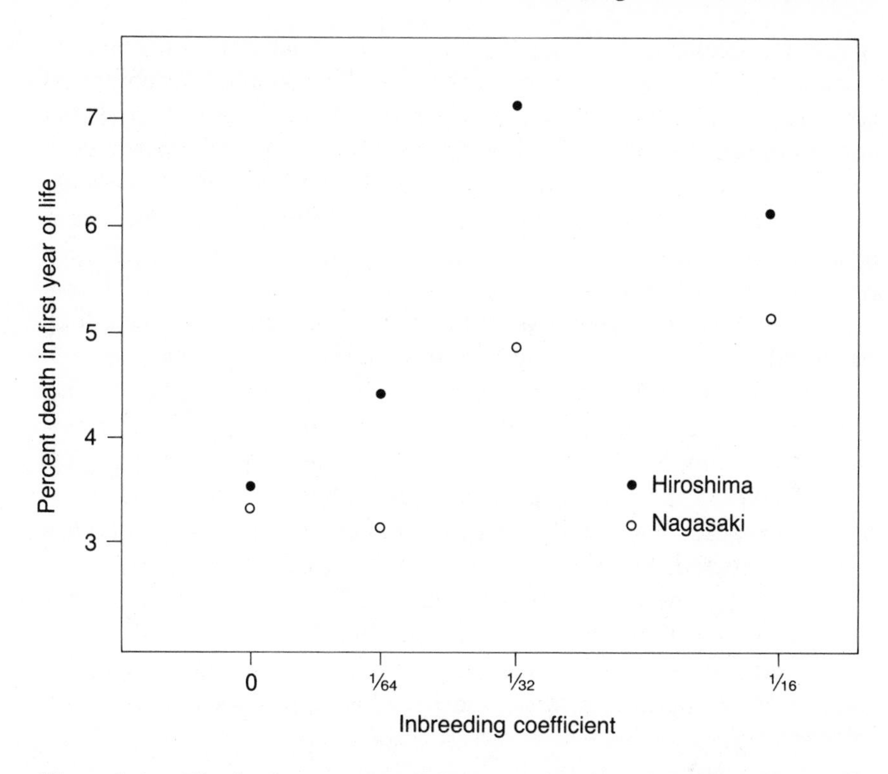

Figure 5-6. The death rate among infants as a function of the inbreeding coefficient in the populations of Hiroshima and Nagasaki. (Schull and Neel, 1965.)

In an investigation of the parents of diseased infants, 25 percent were found to be cousins.

d. Does this information change your answer to questions (a) and (b)? If so, specify the direction of change.

5.2 Counseling on Mendelian Diseases

For diseases determined by known Mendelian factors, risks of recurrence of the disease within a family are, in principle, easy to compute. In practice, variable penetrance or late onset may cause difficulties. If the disease is to be modeled in one of the ways described so far, then it is important from the patient's point of view to know whether the model is sufficiently supported. As we have pointed out, to distinguish between models is difficult in practice, and many practitioners prefer to use the information from the patient's family together with selected information (e.g., distribution of onset ages) from the general population to compute the recurrence risk. This is called the empirical approach. The following is an example, based on an example by Vogel and Motulsky (1979, p. 596) that uses both types of information.

A healthy woman whose brother has X-linked Duchenne-type muscular dystrophy and whose maternal uncle had the disease, seeks counseling after having one healthy son. She wants to know the risk for her next son to be affected.

The woman's mother must be heterozygous for the disease allele, so initially her chance of being heterozygous is 1/2, in which case each son would have a 50 percent chance of being affected. As she has already had one healthy son, however, her probability of being heterozygous is now 1/3. To get further information on her genotype, her activity level of the enzyme creatinine phosphokinase (CPK) is determined, because in about 70 percent of heterzygotes the CPK activity level is increased. She shows a normal CPK level.

Write N for the information on the unaffected woman and N_s for the information that her first son is healthy. We compute the probability that the woman is a heterozygote H given N and N_s. Using Bayes rule of conditional probabilities, this can be written symbolically as

$$P[H|N \text{ and } N_s] = \frac{P[H]P[N \text{ and } N_s|H]}{P[H]P[N \text{ and } N_s|H] + P[\text{not } H]P[N \text{ and } N_s|\text{not } H]},$$

where $P[A|B]$ is the probability of event A given that event B occurred. Now $P(H) = 1/2$. Given that she is a heterozygote, the events that she has a normal CPK level and her son is unaffected are independent, with probabilities 0.3 and 0.5, respectively, that is, $P[N \text{ and } N_s|H] = P[N|H] \times P[N_s|H] = 0.3 \times 0.5$. Given she is not a heterozygote, then she has a normal CPK level and her son will be unaffected, that is, $P[N \text{ and } N_s|\text{not } H] = 1$. Hence

$$P[H|N \text{ and } N_s] = \frac{0.5[0.3 \times 0.5]}{0.5[0.3 \times 0.5] + 0.5 \times 1} = \frac{3}{23}.$$

The risk that her next son will be affected is then $3/23 \times 1/2 = .0652$.

The attractions of this approach must be tempered with some of the same considerations that apply to the modeling approach. For example, the population group with whom the patient is compared must be appropriate in age and in as many other disease-related variables as possible.

The risk of being affected by a dominant disease with limited penetrance or with variable age of onset may be considered from the same point of view as in the preceding example for an X-linked disease. The problem is again whether a healthy individual carries the disease. If an extended pedigree is available, and if the disease gene is known to be linked to marker loci revealed by serological, biochemical, or restriction fragment length polymorphisms, then risk estimates can be considerably improved. Even when pedigree information is not available, the risk estimates are improved if the disease locus is known to occur in linkage disequilibrium with marker loci. These two kinds of information may aid in counseling for the late onset disease Huntington's chorea, which shows tight linkage to the RFLP defined by the probe *G8* and the restriction enzyme *HindIII* (section 4.2.1). The current rapid development of DNA techniques promises increasing precision in the definition and analysis of marker loci. The ultimate goal is, of course, to uncover a RFLP that is absolutely linked to the disease locus, that is, one that involves a recognition site within the disease gene that distinguishes the disease allele from the normal allele.

Counseling for recessive diseases is fundamentally different from that for X-linked or dominant autosomal diseases. A rare recessive disease is not generally revealed in the passage from parents to offspring; rather, it reveals itself as aggregating within sibships. Typically, even the risk that a healthy individual with an affected sib has a diseased child is small. This risk will be of the order of magnitude of the frequency of the disease allele, which is usually well below 0.01. Notable exceptions are the hemoglobinopathies such as sickle cell anemia, where the gene frequency in, for example, the U.S. black population reaches 0.05.

For groups with a high risk of sickle cell anemia, counseling is easier because the carriers are phenotypically recognizable through the sickle cell trait. Thus, the risk that a couple bears children with sickle cell anemia can be assessed from the phenotypes of the parents. Similar assessments can be made for all the major hemoglobinopathies mentioned in section 1.2.2. The use of RFLPs that occur in linkage disequilibrium with the disease allele may permit the extension of carrier assessment to other recessive diseases.

These methods provide married couples with risk estimates that aid them in deciding whether to bear children. It can relieve the anxiety of some couples and eliminate the need to make decisions about having an otherwise wanted child. In addition, the development of biochemical assays of proteins and restriction enzyme analysis of DNA allows prenatal information on a conceived child's genotype through amniocentesis.

Exercise 5.2.A

Consider a recessive autosomal disease that occurs in a frequency of 1 per 10,000 and that develops late in childhood.

a. What is the probability that the individual marked ? in the following pedigree gets the disease?

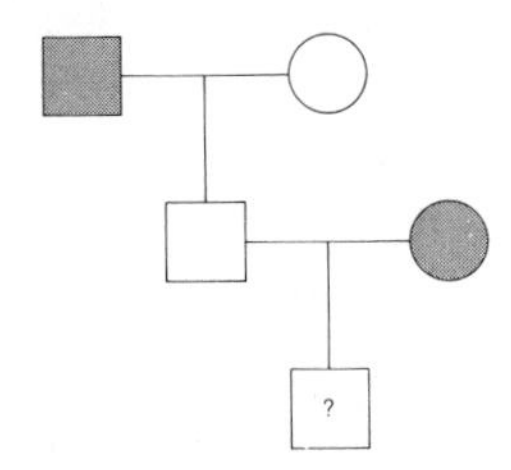

(Shaded individuals have the disease and are homozygous *aa* for the disease allele. Unshaded individuals are normal and carry at least one normal allele, that is, their genotype is either *Aa* or *AA*.)

b. The same question for the pedigree:

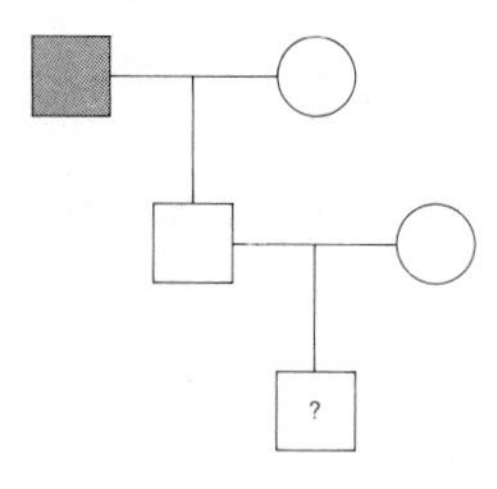

c. And for

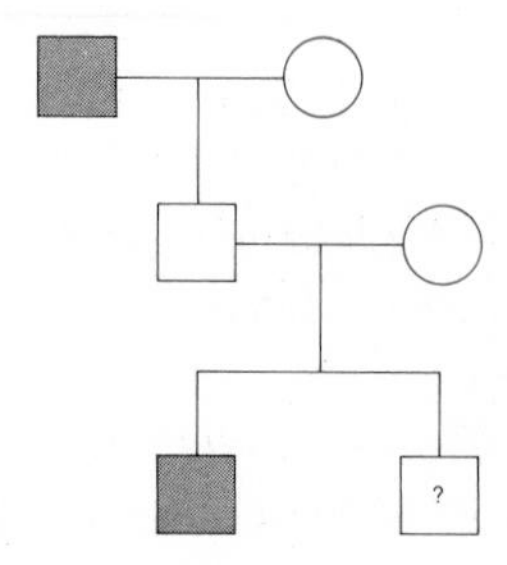

d. And for

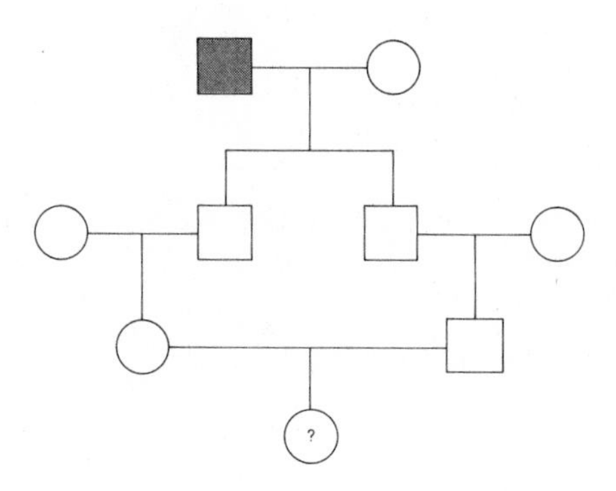

An approximate answer is sufficient.

Suppose now that a marker locus with alleles M_1 and M_2 is known to be at a recombination distance of 5cM (a recombination frequency of 5%) from the disease locus.

e. What is the probability that the individual marked ? in the following pedigree gets the disease when his genotype at the marker locus is M_1M_2?

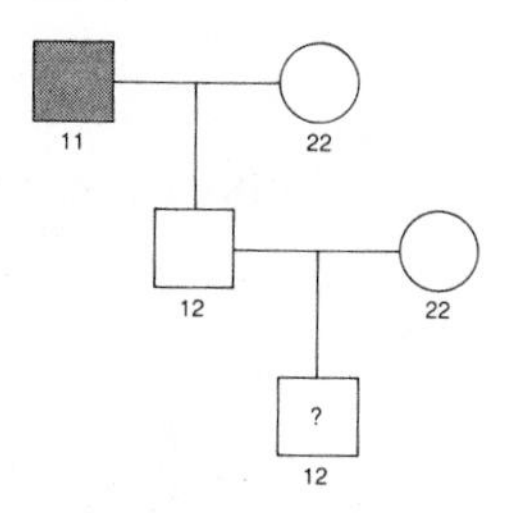

(The genotype at the marker locus is given by the numbers under each individual.)

In the population another marker locus with alleles R_1 and R_2 occurs in frequencies that differ between diseased and normal individuals. The frequency of allele R_1 is 0.8 among individuals with the disease and allele R_1 occurs in a frequency below 0.01 among normal individuals. Recombination has never been observed.

f. What is the probability that the individual marked ? in the following pedigree gets the disease given that he has the genotype R_1R_2?

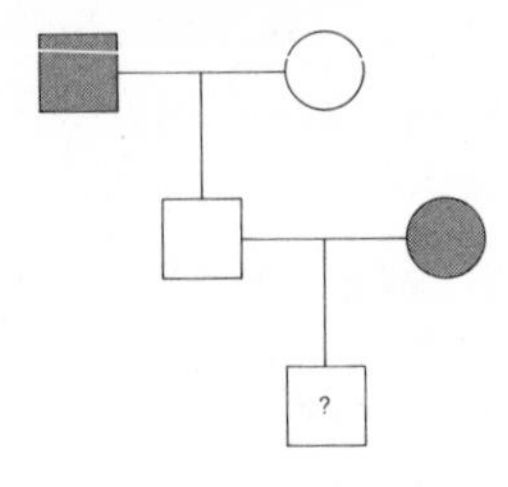

(Answer: about 2/7.)

g. The same question for this pedigree

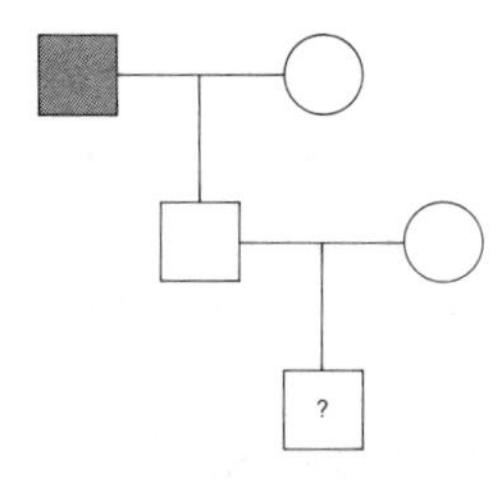

still given that the individual marked ? has the genotype R_1R_2, but assuming now that both of his parents are known to have the genotype R_1R_2 at the marker locus.
(Answer: very small.)

h. The same question, but now the father has genotype R_2R_2 and his mother R_1R_2.
(Answer: about 1/2.)

5.3 Eugenic Consideration of Mendelian Diseases

Eugenics is the study of manipulation of the human germ plasm to improve the human species. Within this context the possibility of lowering the frequency of Mendelian genetic diseases by breeding programs has received much attention. It has become apparent, however, that no feasible procedure can produce any appreciable effect on the disease incidences.

In general, the most serious genetic diseases cause the diseased person to have a very low fitness; lowering that fitness (e.g., by sterilization) will not produce any significant effect on the disease incidence either in the short or in the long run. Further, such a procedure will have virtually no effect on recessive diseases, as only a trivial fraction of the disease alleles are expressed phenotypically. Interference with the parents of a diseased child will have a similarly low effect on the allele frequency. Many such parents will pursue genetic counseling and medical help, but eugenic considerations are irrelevant in providing services in these instances.

Many Mendelian genetic diseases can be treated medically to ameliorate the symptoms of the disease. This will often increase the population genetic fitness of the diseased individuals. For dominant diseases this will produce an immediate increase in the disease incidence by a factor s/s', where s' is the selection coefficient of the treated individuals. For recessive diseases a similar increase in the equilibrium disease incidence is expected, but no effect on the incidence will be noticeable for hundreds of generations.

Prenatal diagnosis of diseased fetuses combined with induced abortions of these fetuses have a special effect. Abortions can be compensated for by an increased number of pregnancies, with the effect that marriages at risk have an effective fecundity higher than the population average. For dominant diseases this does not alter the qualitative properties of the population dynamics, but for recessive diseases the increased fecundity is imparted to the heterozygote by heterozygote marriages. Thus, the net effect is an increased fitness of heterozygotes, which qualitatively resembles the situation of malaria resistance (sections 1.2.2 and 3.2.2). The result of fetal selection against the diseased homozygote may then be that the gene frequency of the disease allele increases in the population to a stable equilibrium determined by the amount of compensation in fecundity of marriages at risk.

The incidence of all Mendelian genetic diseases is proportional to the mutation rate. Therefore, an environmentally induced increase in the mutation rate will produce a similar increase in the frequency of genetic diseases. For dominant diseases this effect will be seen immediately after the environmental change, and a doubling of the mutation rate will be seen within a few generations as a doubling of the disease incidence. For recessive diseases a doubling of the mutation rate is expected to produce a doubling in the equilibrium disease rate, but this will be attained extremely slowly. The largest change will be seen in the first generation after the mutation increase, and the largest effect in a randomly mating population is that $2(\mu' - \mu)q$ will be added to the previous disease incidence.

The new techniques of gene transformation through recombinant DNA techniques may promise a treatment of Mendelian genetic diseases. The specific repair of a damaged gene in a relevant tissue may change the present treatment of symptoms to a cure for genetic diseases. The population genetics effect of this is to further increase the fitness of the diseased individual. As long as the gene transformations are performed in somatic tissues and the germ line cells are untouched, however, the treatment will have no direct genetic effects.

The ultimate cure for a disease is, of course, to halt its transmission and transform the damaged gene in the germ line cells of the otherwise cured individual. Although this is at present a subject of science fiction, we may consider such a hypothetical gene cure from a population genetics point of view. Any change in the germ line cells amounts to a mutation, and a gene cure would be a desirable mutation from a damaged gene to a fully functional normal gene. If this prerequisite is not fulfilled, it is not a complete cure. In the future design of such a cure, however, one should look carefully for side effects. The induced mutation should be restricted specifically to the gene needing a repair. It would be a bad cure if its cost were mutations in other genes, mutations that have a high probability of being harmful. In any eugenic consideration of simple genetic diseases, the mutation-selection balance equation of section 5.1.2 should be kept in mind. Any harmful mutation will stay in the population until it is eliminated by selection, that is, until it is eliminated because a diseased person failed to survive and reproduce.

Answers to Selected Exercises

Answer: 1.1.B

Location 1: $\hat{m} = 107.98$, $\hat{V} = 3.71$.
Location 3: $\hat{m} = 114.01$, $\hat{V} = 4.90$.

Answer: 2.1.A

a. A_1A_1 mothers transmit only A_1 alleles to the offspring, A_2A_2 mothers transmit only A_2.

b. $q_1 = 41/111$; $q_2 = 70/111$.

c. $q_1 = 127/314$; $q_2 = 187/314$.

d. 1/2. To see this, assume that the allele frequencies among sperm are q_1 for A_1 and q_2 for A_2. Then a heterozygote offspring is formed by an A_1 egg and an A_2 sperm in the frequency $\frac{1}{2} q_2$, and by an A_2 egg and an A_1 sperm in the frequency $\frac{1}{2} q_1$. In total, $(q_1 + q_2)/2 = 1/2$.

e. $q_1 = 65/184$; $q_2 = 119/184$. The frequency of the A_1A_2–A_1A_2 class is the same for all values of q_1 and q_2, so no information on male gametes is contained in this class.

f. A_1: 41 + 65 + 127; A_2: 70 + 119 + 187.

g. $q_1 = 233/609$; $q_2 = 376/609$.

h. $p_1 = (222 + 357)/1564$; $p_2 = (357 + 628)/1564$.

i. $p_1' = (212 + 370)/1564$; $p_2' = (370 + 612)/1564$.

j. $(q_1 + p_1)/2 = 0.376$; $p_1' = 0.372$.

The reason for the discrepancy is that the frequency of the heterozygous offspring from heterozygous mothers is not exactly equal to 1/2 but is 0.485 in this data. This small deviation from the expectation is due to the random effects of limited sample size. In fact, the observed deviation is considerably smaller than the largest acceptable deviation in a sample of this size.

Answer: 2.1.B

In the following we give the allele frequencies of A_1 only.

a. Mothers: 579/1564
   Nonbreeding females: 48/138
   Adult females: (579 + 48)/(1564 + 138)
   Adult males: 308/862
   Male gametes: 233/609

b. (579 + 48 + 308 + 233)/(1564 + 138 + 862 + 609).
   The allele count includes mothers, nonbreeding females, adult males, and male gametes. The adult females are just the sum of the two observed female classes, so they add nothing new. Where are the zygotes? Since they were taken randomly, one from each mother, one of their alleles was already observed in the mother. Hence the observation of this allele is not independent of the alleles counted in the mothers. The other allele in the zygotes is counted as observed male gametes, except for those hidden behind the Mendelian segregation law demonstrated in Exercise 2.1.A, question (e).

Answer: 2.2.B

The genotypic frequencies among zygotes are p_1q_1, $p_1q_2 + p_2q_1$, and p_2q_2 for A_1A_1, A_1A_2, and A_2A_2, respectively, where p_1 and p_2 are the allele frequencies among breeding females and q_1 and q_2 those among breeding males. The allele frequencies among zygotes are $p_i' = (p_1 + q_1)/2$ and $p_2' = (p_2 + q_2)/2$. Thus, the frequency of heterozygotes in the corresponding Hardy-Weinberg proportions is $(p_1 + q_1)(p_2 + q_2)/2$.

Now consider the difference

$$(p_1q_2 + p_2q_1) - \frac{(p_1 + q_1)(p_2 + q_2)}{2}.$$

On rearrangement this is

$$\frac{p_1q_2 + p_2q_1 - p_1p_2 - q_1q_2}{2},$$

or

$$\frac{(p_1 - q_1)(q_2 - p_2)}{2}.$$

Remember that $p_2 = 1 - p_1$ and $q_2 = 1 - q_1$, so the difference is actually

$$\frac{(p_1 - q_1)^2}{2},$$

which is always positive when $p_1 \neq q_1$.

For an X-linked locus, this result shows that the frequency of heterozygotes among females is always higher than the frequency given by the Hardy-Weinberg proportions corresponding to the allele frequencies among females, unless the population has reached the equilibrium where the allele frequencies are equal in the two sexes (Figure 2-1).

Answer: 2.2.C

a. The frequency of a given pair of sibs varies among the different possible genotypic combinations of their parents. Let us first evaluate these frequencies of sib combinations within matings:

Mating type	Frequency	Frequency of sib genotypes within mating									
		A_1A_1			A_1A_2			A_2A_2			Total
		A_1A_1	A_1A_2	A_2A_2	A_1A_1	A_1A_2	A_2A_2	A_1A_1	A_1A_2	A_2A_2	
$A_1A_1 \times A_1A_1$	p_1^4	1	0	0	0	0	0	0	0	0	1
$A_1A_1 \times A_1A_2$	$4p_1^3p_2$	$\frac{1}{4}$	$\frac{1}{4}$	0	$\frac{1}{4}$	$\frac{1}{4}$	0	0	0	0	1
$A_1A_1 \times A_2A_2$	$2p_1^2p_2^2$	0	0	0	0	1	0	0	0	0	1
$A_1A_2 \times A_1A_2$	$4p_1^2p_2^2$	$\frac{1}{16}$	$\frac{1}{8}$	$\frac{1}{16}$	$\frac{1}{8}$	$\frac{1}{4}$	$\frac{1}{8}$	$\frac{1}{16}$	$\frac{1}{8}$	$\frac{1}{16}$	1
$A_1A_2 \times A_2A_2$	$4p_1p_2^3$	0	0	0	0	$\frac{1}{4}$	$\frac{1}{4}$	0	$\frac{1}{4}$	$\frac{1}{4}$	1
$A_2A_2 \times A_2A_2$	p_2^4	0	0	0	0	0	0	0	0	1	1

From this table and the frequencies of the various matings, the answer becomes:

	A_1A_1	A_1A_2	A_2A_2	Total
A_1A_1	$\frac{1}{4}p_1^2(1+p_1)^2$	$\frac{1}{2}p_1^2p_2(1+p_1)$	$\frac{1}{4}p_1^2p_2^2$	p_1^2
A_1A_2	$\frac{1}{2}p_1^2p_2(1+p_1)$	$p_1p_2(1+p_1p_2)$	$\frac{1}{2}p_1p_2^2(1+p_2)$	$2p_1p_2$
A_2A_2	$\frac{1}{4}p_1^2p_2^2$	$\frac{1}{2}p_1p_2^2(1+p_2)$	$\frac{1}{4}p_2^2(1+p_2)^2$	p_2^2

b. The genotypic proportions among the firstborn children are, as expected, the Hardy-Weinberg proportions.

A quicker way to solve this problem after reading section 2.4.1 is to use Cotterman's coefficients to describe the identity relationships between the two sibs as follows. The two genes in the second sib are identical with the two genes in the first sib with probability of 1/4. In this event, the genotypes of the sibs are necessarily the same, no matter what the genotypes of the parents. Similarly, the two genes in the second sib are both nonidentical to the genes in the first sib, with probability 1/4. In this event, the genotypic proportions among second sibs are just like the genotypic proportions of a random individual, that is, the Hardy-Weinberg proportions, if nothing is known about the parents. The second sib has one of its genes identical to one of the genes in the first sib, with probability 1/2. In this event, the other gene of the second sib is allele A_1 with probability p_1 and allele A_2 with probability p_2. Thus the table becomes

	A_1A_1	A_1A_2	A_2A_2
A_1A_1	$p_1^2\left[\frac{1}{4}+\frac{1}{4}p_1^2+\frac{1}{2}p_1\right]$	$p_1^2\left[\frac{1}{4}(2p_1p_2)+\frac{1}{2}p_2\right]$	$p_1^2\left[\frac{1}{4}p_2^2\right]$
A_1A_2	$2p_1p_2\left[\frac{1}{4}p_1^2+\frac{1}{2}\left(\frac{1}{2}p_1\right)\right]$	$2p_1p_2\left[\frac{1}{4}+\frac{1}{4}(2p_1p_2)+\frac{1}{2}\left(\frac{1}{2}p_1+\frac{1}{2}p_2\right)\right]$	$2p_1p_2\left[\frac{1}{4}p_2^2+\frac{1}{2}\left(\frac{1}{2}p_2\right)\right]$
A_2A_2	$p_2^2\left[\frac{1}{4}p_1^2\right]$	$p_2^2\left[\frac{1}{4}(2p_1p_2)+\frac{1}{2}p_1\right]$	$p_2^2\left[\frac{1}{4}+\frac{1}{4}p_2^2+\frac{1}{2}p_2\right]$

Answer: 2.3.D

a. Write the gamete frequencies as

$$x_1 = p_{A_1}p_{B_1} + D,$$

$$x_2 = p_{A_1}p_{B_2} - D,$$

$$x_3 = p_{A_2}p_{B_1} - D,$$

$$x_4 = p_{A_2}p_{B_2} + D.$$

As the gamete frequencies cannot be negative, we see that

$$D > -p_{A_1}p_{B_1}$$

$$D < p_{A_1}p_{B_2}$$

$$D < p_{A_2}p_{B_1}$$

$$D > -p_{A_2}p_{B_2}$$

Therefore, if D is negative, then $|D|$ is less than the smaller of $p_{A_1}p_{B_1}$, and $p_{A_2}p_{B_2}$, and if D is positive, it is less than the smaller of $p_{A_1}p_{B_2}$ and $p_{A_2}p_{B_1}$.

b. Assume that $D < 0$ (implying that $\delta < 0$) and that $p_{A_1}p_{B_1} < p_{A_2}p_{B_2}$. Then from the foregoing,

$$-D < p_{A_1}p_{B_1}.$$

But

$$(p_{A_1}p_{B_1})^2 < (p_{A_1}p_{B_1})(p_{A_2}p_{B_2}),$$

by the assumption, so $-\delta < 1$ and therefore $-1 < \delta$.
If $p_{A_1}p_{B_1} > p_{A_2}p_{B_2}$, a similar argument works. If $D > 0$, then $\delta > 0$, and similar partitioning into the cases $p_{A_1}p_{B_2} < p_{A_2}p_{B_1}$, and $p_{A_2}p_{B_1} < p_{A_1}p_{B_2}$ produces the result $\delta < 1$.

Answer 2.4.C

a. $g_{2s} = \dfrac{1}{2N_1} + \left(1 - \dfrac{1}{2N_1}\right)g_{2s-1}$

b. $g_{2s-1} = \dfrac{1}{2N_2} + \left(1 - \dfrac{1}{2N_2}\right)g_{2s-2}$

c. $(1 - g_{2s}) = \left(1 - \dfrac{1}{2N_1}\right)\left(1 - \dfrac{1}{2N_2}\right)(1 - g_{2s-2})$

d. $\left(1 - \dfrac{1}{2N_e}\right)^2 = \left(1 - \dfrac{1}{2N_1}\right)\left(1 - \dfrac{1}{2N_2}\right)$

so that

$$\frac{1}{N_e} - \frac{1}{4}\left(\frac{1}{N_e}\right)^2 = \frac{1}{2}\left(\frac{1}{N_1} + \frac{1}{N_2}\right) - \frac{1}{4N_1N_2}$$

This equation can be solved for N_e, but if either of N_1 or N_2 are reasonably large, then approximately

$$N_e = \left[\frac{1}{2}\left(\frac{1}{N_1} + \frac{1}{N_2}\right)\right]^{-1},$$

the harmonic mean of N_1 and N_2. If the population size varies in cycles of n generations $N_1, N_2, \ldots, N_n, N_1, N_2, \ldots$, then

$$N_e \approx \left[\frac{1}{n}\sum_{i=1}^{n}\frac{1}{N_i}\right]^{-1}$$

e.

case	N_e	Approximation from (d)
1	12.0	12.0
2	16.6	16.7
3	18.0	18.2
4	19.7	19.8

Answer: 3.1.A

The frequency of heterozygotes is $2pq = 2p(1 - p)$. Differentiating this with respect to p yields $2(1 - p) - 2p = 2(1 - 2p)$. This differential coefficient is positive in the interval $0 < p < 1/2$ and negative in $1/2 < p < 1$. Thus, the frequency of heterozygotes increases when p moves from 0 to 1/2, and decreases from 1/2 to 1. The maximum is therefore reached at $p = 1/2$ where the frequency of heterozygotes is 1/2.

Answer 3.2.A

a. $v_{D_1D_1} = \frac{36}{21}, v_{D_1D_2} = 1, v_{D_2D_2} = 1$

b. D_1 will increase to fixation while D_2 is eliminated.

Answer: 4.1.A

The Hardy-Weinberg proportions are $\frac{1}{16}, \frac{6}{16}$ and $\frac{9}{16}$. Thus

$$m = 16 \times \frac{1}{16} + 48 \times \frac{6}{16} + 80 \times \frac{9}{16} = 64,$$

and the genotypic effects are

A_1A_1	A_1A_2	A_2A_2
-48	-16	$+16$

The genetic variance is then

$$V_G = \frac{1}{16} \times 48^2 + \frac{6}{16} \times 16^2 + \frac{9}{16} \times 16^2 = 384.$$

The average allelic effects are given by

$$\alpha_1 = -48 \times \frac{1}{4} - 16 \times \frac{3}{4} = -24$$

$$\alpha_2 = -16 \times \frac{1}{4} + 16 \times \frac{3}{4} = 8,$$

so the additive variance is

$$V_A = 2\left(\frac{1}{4} \times 24^2 + \frac{3}{4} \times 8^2\right) = 384.$$

Since $V_G = V_A$, we have $V_D = 0$.
Note the additivity of the genotypic values:

$$\frac{1}{2} \times 16 + \frac{1}{2} \times 80 = 48.$$

Try to show that in general, additive genotypic values such as the following

A_1A_1	A_1A_2	A_2A_2
$2a_1$	$a_1 + a_2$	$2a_2$

produce $V_D = 0$ for any allele frequencies, p_1 and p_2.

References

Allison, A. C. 1954. The distribution of the sickle-cell trait in East Africa and elsewhere, and its apparent relationship to the incidence of subtertian malaria. *Trans. R. Soc. Trop. Med. Hyg.* 48: 312–318.

André, J. 1973. Recherches écologiques sur les populations de *Cepaea nemoralis* du Languedoc et du Rousillon. Ph.D. thesis. Univ. Paris.

Ayala, F. J., M. L. Tracey, L. G. Barr, J. F. McDonald, and S. Perez-Salas. 1974. Genetic variation in natural populations of five *Drosophila* species and the hypothesis of the selective neutrality of protein polymorphisms. *Genetics* 77: 343–384.

Bennett, P. H., N. B. Rudiforth, M. Miller, and P. M. LeCompte. 1976. Epidemiologic studies of diabetes in the Pima Indians. In *Recent Progress in Hormone Research,* vol. 32, pp. 333–371. Academic Press, New York.

Bodmer, W. F., and J. G. Bodmer. 1978. Evolution and function of the HLA system. *British Medical Bulletin* 34: 309–316.

Bulmer, M. G. 1970. *The Biology of Twinning in Man.* Clarendon Press, Oxford.

Burks, B. S. 1928. The relative influence of nature and nurture upon mental development: A comparative study of foster parent–foster child resemblance and true parent–true child resemblance. *Yearbook of the National Society for the Study of Education* 27: 219–316.

Cahill, G. F., Jr. 1979. Current concepts of diabetic complications with emphasis on hereditary factors: a brief review. In *Genetic Analysis of Common Diseases: Application to Predictive Factors in Coronary Disease,* eds. C. F. Sing and M. Skolnick, pp. 113–125. Alan R. Liss, New York.

Cain, A. J., and P. M. Sheppard. 1954. Natural selection in *Cepaea. Genetics* 39: 89–116.

Carter, C. O. 1965. The inheritance of common congenital malformations. *Prog. Med. Genet.* 4: 59–84.

Cavalli-Sforza, L. L., and W. F. Bodmer. 1971. *The Genetics of Human Populations.* W. H. Freeman, San Francisco.

Cavalli-Sforza, L. L., and A. W. F. Edwards. 1963. Analysis of human evolution. In *Genetics Today,* vol. 3, ed. S. J. Geerts, *Proceedings of the Eleventh International Congress of Genetics.* Pergamon Press, New York.

Cavalli-Sforza, L. L., and M. W. Feldman. 1978. The evolution of continuous variation III. Joint transmission of genotype, phenotype and environment. *Genetics* 90: 391–425.

Christiansen, F. B. 1977. Population genetics of *Zoarces viviparus* (L.): a review. In *Measuring Selection in Natural Populations,* eds. F. B. Christiansen and T. M. Fenchel, pp. 21–47. *Lecture Notes in Biomathematics,* vol. 19. Springer Verlag, Berlin.

Christiansen, F. B., and O. Frydenberg. 1973. Selection component analysis of natural polymorphisms using population samples including mother-offspring combinations. *Theor. Pop. Biol.* 4: 425–445.

Christiansen, F. B., and O. Frydenberg. 1974. Geographical patterns of four polymorphisms in *Zoarces viviparus* as evidence of selection. *Genetics* 77: 765–770.

Christiansen, F. B., and O. Frydenberg. 1977. Selection-mutation balance for two non-allelic recessives producing an inferior double homozygote. *Amer. J. Hum. Genet.* 29: 195–207.

Christiansen, F. B., O. Frydenberg, J. P. Hjorth, and V. Simonsen. 1976. Genetics of *Zoarces* populations IX. Geographic variation at the three phosphoglucomutase loci. *Hereditas* 83: 245–256.

Christiansen, F. B., O. Frydenberg, and V. Simonsen. 1973. Genetics of *Zoarces* populations IV. Selection component analysis of an esterase polymorphism using population samples including mother-offspring combinations. *Hereditas* 73: 291–304.

Cooper, D. N., and J. Schmidtke. 1984. DNA restriction fragment length polymorphisms and heterozygosity in the human genome. *Hum. Genet.* 66: 1–16.

Darwin, C. 1862. On the two forms, or dimorphic condition, in the species of *Primula,* and on their remarkable sexual relations. *J. Proc. Linnean Soc. (Botany)* 6: 77–96.

Dausset, J., and A. Svejgaard, eds. 1977. *HLA and Disease.* Munksgaard, Copenhagen.

Dobzhansky, T, and H. Levene. 1951. Development of heterosis through natural selection in experimental populations of *Drosophila pseudoobscura. Amer. Natur.* 85: 247–264.

Ehrlich, P. R., R. W. Holm, and D. R. Parnell. 1974. *The Process of Evolution.* McGraw-Hill, New York.

Elandt-Johnson, R. C. 1971. *Probability Models and Statistical Methods in Genetics.* John Wiley & Sons, New York.

Elston, R. C. 1979. Likelihood models in human quantitative genetics. In *Genetic Analysis of Common Diseases: Application to Predictive Factors in Coronary Disease,* eds. C. F. Sing and M. Skolnick, pp. 391–405. Alan R. Liss, New York.

Ewens, W. J. 1979. *Mathematical Population Genetics.* Springer Verlag, Berlin.

Fabricius-Hansen, V. 1939. Blood groups and M-N types of Eskimos in East Greenland. *J. Immunol.* 36: 523–530.

Fabricius-Hansen, V. 1940. Blodtype-bestemmelser i Julianehaab Syddistrikt (ABO- og MN-systemet). *Nord. Med.* 5: 497–499.

Fain, P. R. 1978. Characteristics of simple sibship variance tests for the detection of major loci and application to height, weight and spatial performance. *Ann. Hum. Genet.* 42: 109–120.

Falconer, D. S. 1965. The inheritance of liability to certain diseases, estimated from the incidence among relatives. *Ann. Hum. Genet.* 29: 51–76.

Falconer, D. S. 1981. *Introduction to Quantitative Genetics*. 2nd ed. Longman, London.

Feinleib, M., and R. J. Garrison. 1979. The contribution of family studies to the partitioning of population variation of blood pressure. In *Genetic Analysis of Common Diseases: Application to Predictive Factors in Coronary Disease*, eds. C. F. Sing and M. Skolnick, pp. 653–673. Alan R. Liss, New York.

Feldman, M. W., and L. L. Cavalli-Sforza. 1979a. On hereditary transmission of diseases of complex etiology. In *Genetic Analysis of Common Diseases: Application to Predictive Factors in Coronary Disease*, eds. C. F. Sing and M. Skolnick, pp. 203–227. Alan R. Liss, New York.

Feldman, M. W., and L. L. Cavalli-Sforza. 1979b. Aspects of variance and covariance analysis with cultural inheritance. *Theor. Pop. Biol.* 15: 276–307.

Feldman, M. W., and F. B. Christiansen. 1975. The effect of population subdivision on two loci without selection. *Genet. Res. Camb.* 24: 151–162.

Feldman, W. W., and F. B. Christiansen. 1984. Population genetic theory of the cost of inbreeding. *Amer. Natur.* 123: 642–653.

Feldman, M. W., F. B. Christiansen, and U. Liberman. 1983. On some models of fertility selection. *Genetics* 105: 1003–1010.

Feldman, M. W., and R. C. Lewontin. 1975. The heritability hangup. *Science* 190: 1163–1168.

Fisher, R. A. 1918. The correlation between relatives on the supposition of Mendelian inheritance. *Trans. Roy. Soc. (Edinburgh)* 52: 399–433.

Fitch, W. M., and E. Margoliash. 1967. Construction of phylogenetic trees. *Science* 155: 279–284.

Frydenberg, O. 1964. Population studies of a lethal mutant in *Drosophila melanogaster* II. Behavior in populations with overlapping generations. *Hereditas* 51: 31–63.

Frydenberg, O., A. O. Gyldenholm, J. P. Hjorth, and V. Simonsen. 1973. Genetics of *Zoarces* populations III. Geographic variations in the esterase polymorphism *EstIII*. *Hereditas* 73: 233–238.

Goldberger, A. S. 1978a. Pitfalls in the resolution of IQ inheritance. In *Genetic Epidemiology*, eds. N. E. Morton and C. S. Chung. Academic Press, New York.

Goldberger, A. S. 1978b. Models and methods in the IQ debate. Part I. Revised. University of Wisconsin Social Systems Research Institute.

Graham, S. M., W. B. Watt, and L. F. Gall. 1980. Metabolic resource allocation vs. mating attractiveness: Adaptive pressures on "alba" polymorphism of *Colias* butterflies. *Proc. Natl. Acad. Sci. U.S.A.* 77: 3615–3619.

Gürtler, H., and K. Henningsen. 1954. Rh-chromosome frequencies in the Danish population. *Acta Path. Microbiol. Scand.* 34: 493–496.

Gusella, J. F., N. S. Wexler, P. M. Conneally, S. L. Naylor, M. A. Anderson, R. E. Tanzi, P. C. Watkins, K. Ottina, M. R. Wallace, A. Y. Sakaguchi, A. B. Young, I. Shoulson, E. Bonilla, and J. B. Martin. 1983. A polymorphic DNA marker genetically linked to Huntington's disease. *Nature* 306: 234–238.

Harburg, E., M. A. Schork, J. C. Erfurt, W. J. Schull, and C. Chape. 1977. Heredity, stress and blood pressure, a family set method II. *J. Chron. Dis.* 30: 649–658.

Harris, H. 1966. Enzyme polymorphisms in man. *Proc. Roy. Soc.* ser. B, 164: 298–310.

Harris, H. 1970. *The Principles of Human Biochemical Genetics*. Elsevier and North-Holland Biomedical Press, Amsterdam.

Hjorth, J. P., and V. Simonsen. 1975. Genetics of *Zoarces* populations VIII. Geographic variation common to the polymorphic loci HbI and EstIII. *Hereditas* 81: 173-184.

Hummel, K., G. Pulverer, K. P. Schaal, and V. Weidtman. 1970. Häufigkeit der Sichttypen in den Erbsystemen Haptoglobin, Gc, saure Erythrocyten-phosphatase, Phosphoglucomutase und Adenylatkinase sowie den Erbeigenschaften Gm(1), GM(2), und Inv(1) bei Deutschen (aus dem Raum Freiburg i. Br. und Köln) und bei Türken. *Humangenetik* 8: 330–333.

Huntley, R. M. C. 1966. Heritability of intelligence. In *Genetic and Environmental Factors in Human Ability*, eds. J. E. Meade and A. S. Parkes, pp. 201–218. Oliver and Boyd, Edinburgh.

Johannsen, W. 1903. Über Erblichkeit in Populationen und in Reinen Linien. G. Fisher, Jena.

Jones, J. S., B. H. Leith, and P. Rawlings. 1977. Polymorphism in Cepaea: a problem with too many solutions? *Ann Rev. Ecol. Syst.* 8: 109–143.

Kamin, L. J. 1974. *The Science and Politics of IQ*. Erlbaum, Hillsdale, N.J.

Kan, Y. W., and A. M. Dozy. 1978. Polymorphism of DNA sequence adjacent to human β-globin structural gene. Relationship to sickle mutation. *Proc. Natl. Acad. Sci. U.S.A.* 75: 5631–5635.

Karlin, S. 1968. *Equilibrium Behavior of Population Genetic Models with Nonrandom Mating*. Gordon and Breach, London.

Karlin, S. 1979. Models of multifactorial inheritance: I, Multivariate formulations and basic convergence results. *Theor. Pop. Biol.* 15: 308–355.

Karlin, S., and J. McGregor. 1972. Polymorphisms for genetic and ecological systems with the weak coupling. *Theor. Pop. Biol.* 3: 210–238.

Karlin, S., and P. T. Williams. 1981. Structured exploratory data analysis (SEDA) for determining mode of inheritance of quantitative traits. II. Simulation studies on the effect of ascertaining families through high-valued probands. *Am. J. Hum. Genet.* 33: 282–292.

Karlin, S., P. T. Williams, and D. Carmelli. 1981. Structured exploratory data analysis (SEDA) for determining mode of inheritance of quantitative traits. I. Simulation studies on the effect of background distributions. *Am. J. Hum. Genet.* 33: 262–281.

Kettlewell, H. B. D. 1973. *The Evolution of Melanism*. Clarendon Press, Oxford.

Kimura, M. 1955. Solution of a process of random genetic drift with a continuous model. *Proc. Natl. Acad. Sci. U.S.A.* 41: 144–150.

Kimura, M. 1983. *The Neutral Theory of Molecular Evolution*. Cambridge University Press, Cambridge.

Kimura, M., and T. Ohta. 1971. *Theoretical Aspects of Population Genetics*. Princeton University Press, Princeton, N.J.

Kimura, M., and G. H. Weiss. 1964. The steppingstone model of population structure and the correlation with distance. *Genetics*. 49: 561–576.

King, M. C., R. C. P. Go, R. C. Elston, H. T. Lynch, and N. L. Petrakis. 1980. Allele increasing susceptibility to human breast cancer may be linked to the glutamate-pyruvate transaminase locus. *Science* 208: 406–408.

Koehn, R. K., and W. F. Eanes. 1978. Molecular structure and protein variation within and among populations. *Evol. Biol.* 11: 39–100.

Kravitz, K., M. Skolnick, C. Cannings, D. Carmelli, B. Baty, B. Amos, A. Johnson, N. Mendell, C. Edwards, and G. Cartwright. 1979. Genetic linkage between hereditary hemochromatosis and HLA. *Amer. J. Hum. Genet.* 31: 601–609.

Kreitman, M. 1983. Nucleotide polymorphism at the alcohol dehydrogenase locus of *Drosophila melanogaster*. *Nature* 304: 412–417.

Langley, C. H., E. A. Montgomery, and W. F. Quattlebaum. 1982. Restriction map variation in the Adh region of *Drosophila melanogaster*. *Proc. Natl. Acad. Sci. U.S.A.* 79: 5631–5635.

Leahy, A. M. 1935. Nature-nurture and intelligence. *Genetic Psychology Monographs* 17: 235–308.

Lees, D. R. 1981. Industrial melanism: Genetic adaptation of animals to air pollution. In *Genetic Consequences of Man Made Changes*, eds. J. A. Bishop and L. M. Cook, pp. 129–176. Academic Press, London.

Lewontin, R. C., and J. L. Hubby. 1966. A molecular approach to the study of genic heterozygosity in natural populations. II. Amount of variation and degree of heterozygosity in natural populations of *Drosophila pseudoobscura*. *Genetics* 54: 595–609.

Li, C. C. 1975. Path Analysis—a primer. Boxwood Press, Pacific Grove, Calif.

Luzatto, L., E. A. Usanga, and S. Reddy. 1969. Glucose-6-phosphate dehydrogenase deficient red cells: resistance to infection by malarial parasites. *Science* 164: 839–841.

Malécot, G. 1969. *The Mathematics of Heredity*. W. H. Freeman, San Francisco.

Mather, K. 1964. *Human Diversity*. The Free Press, New York.

McKusick, V. A. 1983. *Medelian Inheritance in Man*. 6th ed. Johns Hopkins University Press, Baltimore.

Mørch, E. T. 1941. *Chondrodystrophic Dwarfs in Denmark*. Munksgaard, Copenhagen.

Morton, N. E., and C. J. Maclean. 1974. Analysis of family resemblance III. Complex segregation analysis of quantitative traits. *Am. J. Hum. Genet.* 26: 489–503.

Motulsky, A. G. 1975. Glucose-6-phosphate dehydrogenase and abnormal hemoglobin polymorphisms—evidence regarding malarial selection. In *The Role of Natural Selection in Human Evolution*, ed. F. M. Salzano, pp. 271–291. North-Holland, Amsterdam.

Mourant, A. F., A. C. Kopec, and K. Domaniewska-Sobczak. 1976. *The Distribution of the Human Blood Groups and Other Polymorphisms*. 2nd ed. Oxford University Press, London.

Nevo, E., A. Beiles, and R. Ben-Shlomo. 1984. The evolutionary significance of genetic diversity: ecological, demographic and life history correlates. In *Evolutionary Dynamics of Genetic Diversity*, ed. G. S. Mani. *Lecture Notes in Mathematical Biology*, vol. 53, pp. 13–213. Springer Verlag, Berlin.

Newman, H. H., F. N. Freeman, and K. J. Holzinger. 1937. *Twins: A Study of Heredity and Environment*. University of Chicago Press, Chicago.

Nilsson-Ehle, H. 1909. Kreuzungsuntersuchungen an Hafer and Weizen. *Lunds Univ. Aarskr. N.F. Atd.*, ser. 2, vol. 5, no. 2: 1–22.

O'Brien, S. J. 1984. *Genetic Maps 1984*. Cold Spring Harbor Laboratory, New York.

O'Donald, P. 1980. *Genetic Models of Sexual Selection*. Cambridge University Press, Cambridge.

O'Donald, P. 1983. *The Arctic Skua. An Account of the Ecology, Genetics and Sociobiology of a Polymorphic Seabird*. Cambridge University Press, Cambridge.

Osborne, R. H., and F. V. DeGeorge. 1959. *Genetic Basis of Morphological Variation*. Harvard University Press, Cambridge.

Pasval, G., D. J. Weatherall, and J. M. Wilson. 1978. Cellular mechanism for the protective effect of haemoglobin S against falciparum malaria. *Nature* 274: 701–703.

Payne, R., M. Feldman, H. Cann, and J. G. Bodmer. 1977. A comparison of HLA data of the North American Black with African Black and North American Caucasoid Populations. *Tissue Antigens* 9: 135–147.

Penrose, L. S. 1952. Measurement of pleiotropic effects in phenylketonuria. *Ann. Eugen.* 16: 134–141.

Race, R. R., and R. Sanger. 1968. *Blood Groups in Man.* 5th ed. F. A. Davis, Philadelphia.

Ramshaw, J. A., J. A. Coyne, and R. C. Lewontin. 1979. The sensitivity of gel electrophoresis as a detector of genetic variation. *Genetics* 93: 1019–1037.

Rao, D. C., N. E. Morton, J. M. Lalonel, and R. Lew. 1982. Path analysis under generalized assortative mating. II. American IQ. *Genet. Res. Camb.* 39: 187–198.

Rao, D. C., N. E. Morton, and S. Yee. 1976. Resolution of cultural and biological inheritance by path analysis. *Amer. J. Hum. Genet.* 28: 228–242.

Rasmuson, M. 1952. Variation in bristle number of *Drosophila melanogaster. Acta Zoolog., Stockholm* 33: 277–307.

Rice, J., C. R. Cloninger, and T. Reich. 1980. The analysis of behavioral traits in the presence of cultural transmission and assortative mating: application to IQ and SES. *Behavior Genetics* 10: 73–92.

Schmidt, J. 1917. Racial investigations I. *Zoarces viviporus* L. and local races of the same. *C.R. Trav. Lab. Carlsberg* 13(3): 277–397.

Schork, M. A., W. J. Schull, E. Harburg, P. Roeper, and C. Chape. 1977. Heredity, stress and blood pressure, a family set method IV. *J. Chron. Dis.* 30: 671–682.

Schrott, H. G., K. A. Bucher, W. R. Clarke, and R. M. Lauer. 1979. The Muscatine hyperlipidemia family study program. In *Genetic Analysis of Common Diseases: Application to Predictive Factors in Coronary Disease,* eds. C. F. Sing and M. Skolnick, pp. 619–646. Alan R. Liss, New York.

Schull, W. J., and J. V. Neel. 1965. *The Effects of Inbreeding on Japanese Children.* Harper & Row, New York.

Sheppard, P. M., and L. M. Cook. 1962. The manifold effects of the *medionigra* gene of the moth *Panaxia dominula* and the maintenance of a polymorphism. *Heredity* 17: 415–426.

Sheridan, A. K., R. Frankham, L. P. Jones, K. A. Rathie, and J. S. F. Barker. 1968. Partitioning of the variance and estimation of genetic parameters for various bristle number characters of *Drosophila melanogaster. Theor. Appl. Genet.* 38: 179–187.

Shields, J. 1962. Monozygotic twins, brought up apart and together. Oxford University Press, London.

Sick, K. 1965. Haemoglobin polymorphism in cod in the Baltic and the Danish Sea. *Hereditas* 54: 19–48.

Simonsen, V., and O. Frydenberg. 1972. Genetics of *Zoarces* populations II. Three loci determining esterase isozymes in eye and brain tissue. *Hereditas* 70: 235–242.

Sing, C. F., and M. Skolnick, ed. 1979. *Genetic Analysis of Common Diseases: Application to Predictive Factors in Coronary Disease.* Alan R. Liss, New York.

Singh, R. S., R. C. Lewontin, and A. A. Felton. 1976. Genic heterogeneity within electrophoretic alleles of xanthine dehydrogenase in *Drosophila pseudoobscura. Genetics* 84: 609–629.

Smith, K. 1922. Racial investigations VI. Statistical investigations on inheritance in *Zoarces viviporus* L. *C.R. Trav. Lab. Carlsberg* 14(11): 1–64.

Sokal, R. R., and F. J. Rohlf. 1981. *Biometry.* 2nd ed. W. H. Freeman, San Francisco.

Spencer, N., P. A. Hopkinson, and H. Harris. 1964. Quantitative differences and gene dosage in the human red cell acid phosphatase polymorphism. *Nature* 201: 299–300.

Sturtevant, A. H., and G. W. Beadle. 1962. *An Introduction to Genetics*. Dover, New York.

Suzuki, D. T., A. J. F. Griffiths, and R. C. Lewontin. 1981. *An Introduction to Genetic Analysis*, 2nd ed. W. H. Freeman, San Francisco.

Tattersall, R. B., and D. A. Pyke. 1972. Diabetes in identical twins. *Lancet* II: 1120–1125.

Terasaki, P. I., ed. 1980. *Histocompatibility Testing*. Proceedings of the Eighth International Histocompatibility Workshop. UCLA Tissue Typing Laboratory, Los Angeles, Calif.

Trimble, B. K., and J. H. Doughty. 1974. The amount of hereditary disease in human populations. *Ann. Hum. Genet.* 38: 199–223.

Vogel, F., and A. G. Motulsky. 1979. *Human Genetics*. Springer Verlag, Berlin.

Waddington, C. H. 1953. Genetic assimilation of an acquired character. *Evolution* 7: 118–126.

Wendt, G. G., and D. Drohm. 1972. *Fortschritte der Allgemeinen und Klinischen Humangenetik*. Vol. 4, *Die Huntingtonsche Chorea*. Thieme, Stuttgart.

Williams, W. R., and D. E. Anderson. 1984. Genetic epidemiology of breast cancer: segregation analysis of 200 Danish pedigrees. *Genetic Epidemiology* 1: 7–20.

Wright, S. 1931. Statistical methods in biology. *J. Amer. Stat. Assoc.* 26: 155–163.

Wright, S. 1937. The distribution of gene frequencies in populations. *Proc. Natl. Acad. Sci. U.S.A.* 23: 307–320.

Wright, S. 1951. The genetical structure of populations. *Ann. Eugen.* 15: 323–354.

Zeleny, C. 1922. The effect of selection for eye facet number in the white bar-eye race of *Drosophila melanogaster*. *Genetics* 7: 1–115.

Index

Index